T0231911

Wavelet Methods for Dynamical Problems

With Application to Metallic, Composite, and Nano-Composite Structures

Wavelet Methods for Dynamical Problems

With Application to Metallic, Composite, and Nano-Composite Structures

S. Gopalakrishnan
Mira Mitra

CRC Press
Taylor & Francis Group
Boca Raton London New York

CRC Press is an imprint of the
Taylor & Francis Group, an **informa** business

CRC Press
Taylor & Francis Group
6000 Broken Sound Parkway NW, Suite 300
Boca Raton, FL 33487-2742

First issued in hardback 2019

ISBN-13: 978-1-4398-0461-2 (hbk)

Library of Congress Cataloging-in-Publication Data

Gopalakrishnan, S.
  Wavelet methods for dynamical problems : with application to metallic, composite, and nano-composite structures / S. Gopalakrishnan and Mira Mitra.
      p. cm.
  Includes bibliographical references and index.
  ISBN 978-1-4398-0461-2 (hard back : alk. paper)
  1. Strength of materials--Mathematical models. 2. Wavelets (Mathematics) 3. Inverse problems (Differential equations) I. Mitra, Mira. II. Title.

  TA407.G637 2010
  620.1'120151863--dc22   2009049453

Visit the Taylor & Francis Web site at
http://www.taylorandfrancis.com

and the CRC Press Web site at
http://www.crcpress.com

To My Wife,
Gopalakrishnan

To My Parents & Brother,
Mira

Contents

List of Figures   xi

List of Tables   xix

Preface   xxi

1 Introduction   **1**
   1.1   Solution of structural dynamics problem   2
   1.2   Solution of wave propagation problem   4
   1.3   Objective and outline of the book   6

2 Integral Transform Methods   **9**
   2.1   Laplace transform   9
   2.2   Fourier transform   10
     2.2.1   Continuous Fourier transform   11
     2.2.2   Discrete Fourier transform   13
     2.2.3   Short term Fourier transform   16
   2.3   Wavelet transform   17
     2.3.1   Multi-resolution analysis with wavelets   18
     2.3.2   Daubechies compactly supported wavelets   20
     2.3.3   Discrete wavelet transform   23

3 Structural Dynamics: Introduction and Wavelet Transform   **37**
   3.1   Free vibration of single degree of freedom systems   38
   3.2   Forced vibration of SDOF system   41
   3.3   Harmonic loading   43
     3.3.1   Periodic excitation   45
   3.4   Response to arbitrary loading   46
     3.4.1   Impulse response   47
     3.4.2   Convolution integral method   48
     3.4.3   Direct time integration methods   49
   3.5   Response of SDOF through wavelet transform   50
     3.5.1   Non-periodic boundary condition   53
   3.6   Free vibration of multi-degree of freedom system   57
   3.7   Modal analysis for forced vibration response of MDOF   59
   3.8   Response of MDOF system using wavelet transform   61

4 Wave Propagation: Spectral Analysis   **63**
   4.1   Spectrum and dispersion relations   63
   4.2   Computations of wavenumbers and wave amplitudes   68
          4.2.1   SVD-based technique   69
          4.2.2   PEP-based technique   70
   4.3   Spectral finite element (SFE) method   71
   4.4   FSFE formulation of Timoshenko beam   73
   4.5   FSFE formulation of isotropic plate under in-plane loading   76

5 Wavelet Spectral Finite Element: Time Domain Analysis   **81**
   5.1   Reduction of wave equations for a rod   81
   5.2   Decoupling using eigenvalue analysis   83
   5.3   Wavelet spectral finite element formulation for a rod   84
   5.4   Time domain response of elementary rod under impulse load   87
          5.4.1   Comparison with finite element solution   87
          5.4.2   Advantage over Fourier transform-based spectral fi-
                   nite element   89
   5.5   Reduction of wave equations for Euler-Bernoulli beam . . .   91
   5.6   WSFE formulation for Euler-Bernoulli beam   92
   5.7   Time domain response of Euler-Bernoulli beam under impulse
          load   95
   5.8   Wave propagation in frame structure   98
   5.9   Governing differential wave equations for higher order com-
          posite beam   101
   5.10   WSFE formulation for composite beam   104
   5.11   Time domain response of higher order composite beam . . .   108
          5.11.1   Response to impulse load   109
          5.11.2   Response to modulated pulse load   112

**6 Wavelet Spectral Finite Element: Frequency Domain Analy-
   sis**   **117**
   6.1   Frequency domain analysis: periodic boundary condition . .   117
   6.2   Computation of wavenumbers and wave speeds   120
          6.2.1   Elementary rod   123
          6.2.2   Euler-Bernoulli and Timoshenko beam   123
          6.2.3   Higher order composite beam   127
   6.3   Constraint on time sampling rate   129

**7 Wavelet Spectral Finite Element: Two-Dimensional Struc-
   tures**   **141**
   7.1   Governing differential wave equations for isotropic plate . .   141
   7.2   Reduction of wave equations through temporal approximation 144
   7.3   Reduction of wave equations through spatial approximation   146
          7.3.1   Un-restraint boundary condition   147
          7.3.2   Restraint boundary condition   148

7.4   Wavelet spectral finite element for plate   150
7.5   Wave propagation in isotropic plates   152
    7.5.1   Frequency domain analysis   155
    7.5.2   Time domain response of plate to impulse load . . .   156
    7.5.3   Wave propagation in ply-dropped plate   161
    7.5.4   Wave propagation in folded plate structure   161
7.6   Governing differential wave equations for axisymmetric cylinder   162
7.7   Bessel function solution for axisymmetric cylinder   164
7.8   Wave propagation in isotropic axisymmetric cylinders . . .   165
    7.8.1   Wavenumber computation   166
    7.8.2   Time domain analysis   166

8   Vibration and Wave Propagation in Carbon Nanotubes   183
8.1   Carbon nanotubes: introduction   183
8.2   Axisymmetric shell model of single-walled carbon nanotubes   185
    8.2.1   Frequency domain analysis   185
8.3   Thin shell model of multi-walled carbon nanotubes   192
    8.3.1   Governing differential equations   193
    8.3.2   Spectral finite element formulation   195
8.4   Frequency domain analysis   196
8.5   Time domain analysis   200

9   Vibration and Wave Propagation in Nano-Composites   207
9.1   Introduction: nano-composites   207
9.2   Beam model of MWNT embedded nano-composite   209
9.3   Spectral finite element formulation for MWNT embedded nano-composite   215
9.4   Frequency domain analysis   217
9.5   Time domain analysis   225
    9.5.1   Response to broad-band impulse load   225
    9.5.2   Response to narrow-banded pulse   229
9.6   Shell model of SWNT-polymer nano-composite   229
    9.6.1   Modeling of pull out and partial interfacial shear stress transfer   231
9.7   Time domain analysis   233
    9.7.1   Pull-out of SWNT   233
    9.7.2   Partial interfacial shear stress transfer   235

10 Inverse Problems   241
10.1   Force reconstruction   241
10.2   Numerical examples of impulse force reconstruction   243
10.3   Damage modeling and detection   247
10.4   Modeling of de-lamination in composite beam   248
10.5   Damage detection and de-noising using wavelet analysis . .   251

10.6   Wave propagation in delaminated composite beam and damage detection 252

References   **265**

Index   **273**

List of Figures

2.1   (a) Discretized (16 points) time signal and (b) DFT coefficients of the signal 15

2.2   (a) Haar (D2) scaling function, (b) Haar wavelet, (c) Daubechies D4 scaling function, (d) Daubechies D4 wavelet, (e) Daubechies D12 scaling function, and (f) Daubechies D12 wavelet 25

2.3   Three level forward Mallat's transform 27

2.4   Three level inverse Mallat's transform 27

2.5   (a) Discretized impulse time signal F(t) and (b) details coefficients $d_{j,k}$ using Haar wavelets 29

2.6   Components of $F(t)$ shown in Figure 2.5 at each wavelet level $j = 0$ to 8 33

2.7   Reconstruction of $F(t)$ shown in Figure 2.5 34

2.8   (a) Wavelet map for the impulse load $F(t)$ shown in Figure 2.5; (b) wavelet map plotted up to 125 kHz 35

3.1   (a) SDOF system: undamped spring-mass; (b) free body diagram 38

3.2   (a) Damped SDOF system: spring-mass-damper; (b) free body diagram 40

3.3   Free vibration of SDOF system under different damping conditions 42

3.4   Step load 42

3.5   Response of under-damped SDOF to step loading 43

3.6   Magnification factor $H(\omega)$ 45

3.7   Phase angle ϕ 46

3.8   Unit impulse 47

3.9   Arbitrary loading 48

3.10   Response of a spring-mass-damper system to periodic square loading (inset) 56

3.11   Response of a spring-mass-damper system to transient pulse loading (inset) 56

3.12   (a) A two-DOF spring-mass system; (b) free body diagrams 57

4.1   Timoshenko beam element with nodal displacements and nodal forces 74

4.2   (a) Plate element; (b) nodal displacements and forces 78

5.1   (a) Rod and (b) beam element with nodal forces and displace-
        ments   85
5.2   Impact load and Fourier transform of the load (inset)   87
5.3   Longitudinal tip velocity in rod due to tip impact load simu-
        lated with time interval (a) $\triangle t = 1\ \mu s$, (b) $\triangle t = 2\ \mu s$, and (c)
        $\triangle t = 4\ \mu s$ 88
5.4   Longitudinal tip velocity in rod due to tip impact load for
        time window T_w (a) $T_w = 1024\ \mu s$, (b) $T_w = 2048\ \mu s$, and
        (c) $T_w = 4096\ \mu s$ 90
5.5   Transverse tip velocity in beam due to tip impact load . . .   98
5.6   Transverse tip velocity in beam due to tip impact load for
        time window T_w (a) $T_w = 1024\ \mu s$, (b) $T_w = 2048\ \mu s$, and
        (c) $T_w = 4096\ \mu s$ 99
5.7   2-D frame structure   100
5.8   Transverse velocity at A of 2-D frame in Figure 5.7 due to the
        applied load P   101
5.9   Transverse tip velocity at A of 2-D frame in Figure 5.7 due to
        the applied load P, for time windows T_w (a) $T_w = 1024\ \mu s$,
        (b) $T_w = 2048\ \mu s$, and (c) $T_w = 4096\ \mu s$ 102
5.10   (a) Beam cross-section and the displacements; (b) composite
        beam element with nodal displacements and forces   103
5.11   Tone-burst signal modulated at 200 kHz and its Fourier trans-
        form (inset)   108
5.12   (a) Axial and (b) transverse tip velocities in a $[0_2/90_2]$ graphite-
        epoxy beam due to tip impulse load applied in axial and trans-
        verse directions respectively   110
5.13   (a) Axial and (b) transverse tip velocities in a graphite-epoxy
        beam with different ply orientations due to tip impulse load
        applied in axial and transverse directions respectively   111
5.14   Axial tip velocities in a graphite-epoxy $[0_2/60_2]$ beam due to
        tip impulse load applied in transverse direction   112
5.15   An infinite beam to observe non-dispersive wave modes . . .   113
5.16   (a) Axial and (b) transverse velocities in a graphite-epoxy
        infinite $[0_2/60_2]$ beam due to modulated sinusoidal pulse at
        200 kHz applied in transverse direction   114
5.17   (a) Axial and (b) transverse velocities in a graphite-epoxy
        infinite $[0_2/60_2]$ beam due to modulated sinusoidal pulse at
        200 kHz applied in axial direction   115

6.1   Comparison of ω_j, λ_j, and γ_j for different order (N) of basis,
        (a) real part and (b) imaginary part of γ_j   121
6.2   Comparison of wavenumbers k_f (exact) and k_w (WSFE), for
        different order (N) of basis   122
6.3   Spectrum relation for elementary rod $k_1 h$ (and pulse spectrum
        '$-.-$') for sampling rate (a) $\triangle t = 1\ \mu s$, (b) $\triangle t = 2\ \mu s$ 124

6.4 Spectrum relation for elementary rod $k_1 h$ (and impulse spectrum '$-.-$') for sampling rate (a) $\triangle t = 4$ μs, (b) $\triangle t = 8$ μs, and (c) imaginary part of $k_1 h$ introduced due to non-periodic WSFE solution with $\triangle t = 4$ μs 125

6.5 Dispersion relation C_g/C_0, $C_0 = \sqrt{EA/\rho A}$ for elementary rod plotted up to $f = 0.6 f_{nyq} = 150$ kHz 126

6.6 Spectrum relation for Euler-Bernoulli beam $k_2 h$ and $k_3 h$ (and pulse spectrum '$-.-$') for sampling rate $\triangle t = 1$ μs 127

6.7 (a) Spectrum relation for Euler-Bernoulli beam $k_2 h$ and $k_3 h$ (and impulse spectrum '$-.-$') for sampling rate $\triangle t = 4$ μs and (b) imaginary part of $k_2 h$ and real part of $k_3 h$ introduced due to non-periodic WSFE solution with $\triangle t = 4$ μs 128

6.8 Dispersion relation C_g/C_0, $C_0 = \sqrt{EA/\rho A}$ for Euler-Bernoulli beam plotted up to $f = 0.6 f_{nyq} = 150$ kHz 129

6.9 Spectrum relation for Timoshenko beam $k_2 h$ and $k_3 h$ using (a) $\triangle t = 2$ μs (and pulse spectrum '$-.-$'); (b) $\triangle t = 8$ μs (and impulse spectrum '$-.-$') 130

6.10 Dispersion relation C_g/C_0, $C_0 = \sqrt{EA/\rho A}$ for Timoshenko beam 131

6.11 Spectrum relation for graphite-epoxy $[0_4]$ composite beam . 131

6.12 Dispersion relation C_g/C_0, $C_0 = \sqrt{EA/\rho A}$ for graphite-epoxy $[0_4]$ composite beam plotted up to $f = 0.6 f_{nyq} = 300$kHz . . 132

6.13 Axial velocity measured at **D** (Figure 5.15), due to sinusoidal pulse modulated at (a) 200 kHz and (b) 110 kHz, applied at **C** with $\triangle t = 1$ μs and $L = 0.5$m 134

6.14 Axial velocity measured at **D** (Figure 5.15), due to sinusoidal pulse modulated at (a) 200 kHz and (b) 110 kHz, applied at **C** with $\triangle t = 2$ μs and $L = 0.5$m 135

6.15 Axial tip velocity of a fixed free rod ($L = 0.25$m) due to tip unit impulse load applied in axial direction with (a) $\triangle t = 4$ μs and (b) $\triangle t = 8$ μs 136

6.16 Transverse velocity measured at **D** (Figure 5.15), due to sinusoidal pulse modulated at 200 kHz applied at **C** with $\triangle t = 1$ μs and $L = 0.5$m 137

6.17 Transverse tip velocity of a fixed free Euler-Bernoulli beam ($L = 0.25$m) due to tip unit impulse load applied in transverse direction with $\triangle t = 4$ μs 138

6.18 Transverse velocity measured at **D** (Figure 5.15), due to sinusoidal pulse modulated at 110 kHz applied at **C** with $\triangle t = 2$ μs and $L = 2.0$m (a) FSFE and WSFE, $N = 22$; (b) WSFE, $N = 6$ 139

6.19 Transverse tip velocity of a fixed free Timoshenko beam ($L = 0.25$m) due to tip unit impulse load applied in transverse direction with $\triangle t = 8$ μs 140

7.1   (a) Plate element and (b) nodal forces and displacements . .   142

7.2   Cantilever uniform plate   155

7.3   The (a) real and (b) imaginary parts of the wavenumbers for
       axial wave propagation   157

7.4   The (a) real and (b) imaginary parts of the wavenumbers for
       transverse wave propagation   158

7.5   Axial velocity of free-free cantilever plate (see Figure 7.2) with
       $L_X = 4.0$ m and $L_Y = 0.5$ m as measured at (a) mid and (b)
       quarter points of the free end AB   159

7.6   Axial velocity of fixed-fixed cantilever plate (see Figure 7.2(a))
       with $L_X = 4.0$ m, (a) $L_Y = 0.5$ m and (b) $L_Y = 0.25$ m
       measured at mid-point of free end AB   168

7.7   Transverse velocity of free-free cantilever plate (see Figure 7.2(a))
       with $L_X = 4.0$ m and $L_Y = 0.5$ m as measured at (a) mid
       and (b) quarter points of the free end AB   169

7.8   Snapshots of axial velocities at time instances (a) $T = 250$ μs
       and (b) $T = 375$ μs in a $[0]_8$ cantilever plate due to tip impulse
       load applied in axial direction   170

7.9   Snapshots of transverse velocities at time instances (a) $T =
       500$ μs and (b) $T = 1000$ μs in a $[0]_8$ cantilever plate due to
       tip impulse load applied in transverse direction   171

7.10   (a) Axial and (b) transverse velocities at mid-point of edge
       AB in asymmetric graphite-epoxy laminates due to tip im-
       pulse load applied in transverse and axial directions along AB,
       respectively   172

7.11   (a) Ply drop and (b) folded plates   173

7.12   (a) Axial and (b) transverse velocities of asymmetric $[0_4/90_4]$
       uniform and ply dropped plate   174

7.13   (a) Axial and (b) transverse velocities at mid-point of BC of
       $[0_8]$ folded plate (Figure 7.11(c)) due load applied along edge
       BE and CF in axial and transverse directions, respectively .   175

7.14   Axisymmetric element with nodal displacements and forces .   176

7.15   (a) Aluminum and (b) aluminum-steel hollow cylinders . . .   176

7.16   (a) Radial and (b) axial internal and external pressure . . .   177

7.17   Real parts of wavenumbers at (a) $r = r_i$, (b) $r = (r_i + r_o)/2$,
       and (c) $r = r_o$ 178

7.18   Radial velocities at mid-point along the axial direction at (a)
       $r = r_i$ and (b) $r = r_o$ due to internal radial pressure   179

7.19   Axial velocities at mid-point along the axial direction at (a)
       $r = r_i$ and (b) $r = r_o$ due to internal axial pressure   180

7.20   (a) Radial and (b) axial velocities in bi-material Al-steel cylin-
       der at mid-point along axial direction at $r = r_i$ 181

8.1 Axisymmetric acoustic phonon dispersion relation for a $(10, 10)$ SWNT with $R = 0.678$ nm and $2h = 0.09$ nm ('$-$o$-$' present model, '$*$' continuum model Ref. [15] and '$-$' abinitio model Ref. [103]) 186

8.2 Variation of frequencies of SWNT with (a) wall thickness $2h$ and (b) radius R, for first five radial vibrational modes . . . 188

8.3 Variation of frequencies of SWNT with (a) wall thickness $2h$ and (b) radius R, for first five longitudinal vibrational modes 190

8.4 Variation of frequencies of SWNT with (a) wall thickness $2h$ and (b) radius R, for first five longitudinal radial coupled vibrational modes 191

8.5 Radial wave propagation snapshots at instances (a) 0.5 ps and (b) 1.0 ps 192

8.6 Schematic of the WSFE model of a three-walled carbon nanotube 197

8.7 Normalized wavenumber $(K_\theta R)$ for (a) single-walled, (b) double-walled, and (c) three-walled carbon nanotubes 199

8.8 Group speeds of SWNT with inner radius (a) 0.678 nm and (b) 2.0 nm 200

8.9 Snapshots of radial wave propagation in SWNT with inner radius $R = 2.0$ nm at time instances (a) 150 ps, (b) 200 ps, (c) 250 ps, and (d) 300 ps 202

8.10 Snapshots of circumferential wave propagation in SWNT with inner radius $R = 2.0$ nm at time instances (a) 150 ps, (b) 200 ps, (c) 250 ps, and (d) 300 ps 203

8.11 Natural frequencies of the first five axial modes of a SWNT with varying inner radius R 203

8.12 Natural frequencies of the first five circumferential modes of a SWNT with varying inner radius R 204

8.13 Natural frequencies of the first five radial modes of a SWNT with varying inner radius R 204

8.14 Axial wave propagation in a three-walled carbon nanotube of innermost radius $R = 2.0$ nm 205

8.15 Circumferential wave propagation in a three-walled carbon nanotube of innermost radius $R = 2.0$ nm 205

8.16 Radial wave propagation in a three-walled carbon nanotube of innermost radius $R = 2.0$ nm 206

9.1 (a) Distributed spring model of coupling between the carbon nanotubes; (b) beam cross-section with embedded double wall carbon nanotube 210

9.2 Spectrum relation for DWNT-aluminum matrix beam for (a) Euler-Bernoulli, (b) Timoshenko, and (c) third order layerwise shear deformation models 219

9.3 Dispersion relation for DWNT-aluminum matrix beam for (a) Euler-Bernoulli, (b) Timoshenko, and (c) third order layerwise shear deformation models 220

9.4 Spectrum relation for TWNT embedded beam for (a) no matrix and (b) polymer-matrix 221

9.5 Spectrum relation for TWNT embedded beam for (a) aluminum-matrix and (b) ceramic-matrix 222

9.6 Dispersion relation for TWNT embedded beam for (a) no matrix and (b) polymer-matrix 223

9.7 Dispersion relation for TWNT embedded beam for (a) aluminum-matrix and (b) ceramic-matrix 224

9.8 Transverse tip velocities in DWNT-polymer composite fixed-free beam due to tip impulse load applied in transverse direction (a) time response and (b) FRF 226

9.9 Transverse tip velocities in DWNT-aluminum composite fixed-free beam due to tip impulse load applied in transverse direction (a) time response and (b) FRF 227

9.10 Transverse tip velocities in DWNT-ceramic composite fixed-free beam due to tip impulse load applied in transverse direction (a) time response and (b) FRF 228

9.11 Transverse velocities in a DWNT-aluminum composite infinite beam due to modulated sinusoidal pulse at 6.0 THz applied in transverse direction (a) third order layerwise deformation and (b) Timoshenko beam theories 230

9.12 Axial velocity at the interface for the (a) SWNT and (b) polymer matrix 234

9.13 FRFs of axial velocity at the interface for the SWNT between frequency ranges (a) 0–11 THz and (b) 30–38 THz 236

9.14 FRFs of axial velocity at the interface for the polymer between frequency ranges (a) 0–11 THz and (b) 30–38 THz 237

9.15 Interfacial shear stress for (a) SWNT and (b) polymer matrix 238

9.16 Radial wave velocity at the interface 239

9.17 Axial velocity at the interface for the (a) SWNT and (b) polymer matrix for partial interfacial shear stress transfer 240

10.1 Longitudinal velocity in rod measured at mid-point ($x = 10\,\text{in}$) due axial impact load at tip ($x = 20\,\text{in}$) using two-dimensional FE 244

10.2 Reconstructed impulse load applied to rod using (a) WSFE and (b) Fourier transform-based technique 244

10.3 Transverse velocity in beam measured at mid-point ($x = 5\,\text{in}$) due transverse impact load at tip ($x = 10\,\text{in}$) using 2-D FE . 245

10.4 Reconstruction of impulse load applied to rod (a) response used as input (Figure 10.1 with simulated white noise) and (b) reconstructed impulse load 246

10.5   Reconstructed impulse load applied to beam (a) WSFE and
        (b) Fourier transform-based technique   246

10.6   Reconstructed impulse load applied to 2-D frame (a) WSFE
        and (b) Fourier transform-based technique   247

10.7   (a) Modeling of an embedded de-lamination with base-laminates
        and sub-laminates. Waveguides 1, 2: base laminates and 3, 4:
        sub-laminates; (b) representation of the base-laminates and
        sub-laminates by spectral elements   249

10.8   Fixed-free beam configuration with mid-plane de-lamination   253

10.9   Narrow-banded pulse modulated at 70 kHz in time and fre-
        quency (inset) domain   253

10.10 Transverse tip velocity of fixed-free graphite-epoxy beam due
        to tip impulse load applied in transverse direction with (a)
        de-lamination length $L_d = 20$ mm (validated with 2-D FE),
        (b) different de-lamination lengths $L_d = 10, 20, 30$ mm, (c)
        de-laminations at different heights above centreline and (d)
        different ply-layup $[0]_8, [45]_8, [60]_8$ 255

10.11 Transverse velocities of fixed-free graphite-epoxy beam due
        to narrow-banded load at 50 kHz applied in transverse direc-
        tion (a) undamaged, (b) de-laminated $L_d = 10$ mm, and (c)
        $L_d = 20$ mm   256

10.12 Dispersion relation of graphite-epoxy $[0]_8$ beam   257

10.13 Transverse velocities of fixed-free graphite-epoxy de-laminated
        ($L_d = 10$ mm) beam due to narrow-banded load at 50 kHz ap-
        plied in transverse direction (a) simulated response, (b) noisy
        response, and (c) de-noised response   258

10.14 Transverse velocities of fixed-free graphite-epoxy de-laminated
        ($L_d = 20$ mm) beam due to narrow-banded load at 50 kHz ap-
        plied in transverse direction (a) $L_1 = 20$mm, (b) $L_1 = 30$mm,
        and (c) $L_1 = 40$mm (see Figure 10.8)   259

10.15 Transverse velocities of fixed-free graphite-epoxy beam due to
        narrow-banded load at 70 kHz applied in transverse direction
        (a) undamaged, (b) de-laminated $L_d = 20$ mm and $L_1 = 40$,
        and (c) de-laminated $L_d = 20$ mm and $L_1 = 90$ 261

10.16 Normalized wavelet coefficients of level corresponding to fre-
        quency content 31.25 – 62.5 kHz of undamaged and damaged
        responses (a) Figures 10.15(a) and (b), respectively and (b)
        Figures 10.15(a) and (c), respectively   262

10.17 Transverse velocities of fixed-free graphite-epoxy de-laminated
        beam due to narrow-banded load at 70 kHz applied in trans-
        verse direction, shown in Figure 10.15(b), (a) noisy response
        and (b) de-noised response   263

10.18 Normalized wavelet coefficients of level corresponding to frequency content 31.25 – 62.5 kHz of undamaged, noisy, and de-noised damaged responses (Figures 10.15(a), 10.17(a) and (b), respectively) 264

List of Tables

2.1 Different windowing functions, $G(t)$ used for WFT 17

2.2 Filter coefficients a_k for Daubechies scaling function with $N =$ 4, 6, and 12 23

5.1 CPU time taken by FE with Newmark's time integration and WSFE methods for simulation of longitudinal tip velocity in rod shown in Figure 5.3 89

5.2 Properties of AS4/3501-6 graphite-epoxy beams 109

Preface

Transform methods are historically known for their ability to solve certain difficult ordinary and partial differential equations. The most commonly used transforms for the solution of these are the Fourier and Laplace transforms. Wavelet transforms are new entrants into this area, although they are quite popular with electrical and communication engineers in characterizing and synthesizing the time signals. Through this book, we show the utility of wavelet transforms in structural engineering by addressing problems involving solutions of ordinary and partial differential equations encountered in dynamics-related problems.

Dynamical problems in structural engineering fall under two categories, one involving low frequencies (of the order of few Hz to few hundred Hz), which we call structural dynamics problems, and the other involving very high frequencies (of the order of kHz to Tera Hz), which we call wave propagation problems. Most problems in structural engineering fall under the former category, wherein the response of the entire structural system is characterized using only the first few vibrational modes. The wave propagation is a multimodal phenomenon involving vibrational modes of very high frequencies. In addition, phase information of the propagating signals also play a crucial role in the response estimation. Then, the question one needs to ask is, "Are wave propagation problems relevant in the context of structural engineering?"

Structural design concepts have undergone a sea of change. The design codes have been recently modified with the addition of stringent design constraints, especially with regard to loading and service conditions. Today's design, in view of these stringent conditions, is a lot lighter and more safe. This has led to rapid progress in structural engineering, especially in the area of advanced materials such as nano-composites, structural health monitoring, and active control of vibrations, and all of these problems deal with high frequency excitation. Conventional analysis tools such as finite element cannot handle these problems due to modeling limitations and extensive computational cost. The only alternative to such problems is the method based on transforms.

Spectral Finite Element (SFE) method is one such transform method, which can be a viable alternative to solving problems involving high frequency excitations. In this regard, SFE based on Fourier transforms is quite well known and established. However, it has severe limitations in handling finite structures and specifying non-zero boundary/initial conditions, and thus its utility in solving real world problems involving high frequency excitation is limited.

On the other hand, SFE based on wavelet transforms, which is one of the main themes of this book, removes all these shortcomings and has potential to becoming a sophisticated analysis tool for handling dynamical problems in structural engineering. The aim of this book is to reinforce this idea.

The next question one needs to answer is: What is the need for nano-composites in a book of this nature? Advances in structural engineering included the development of nano-composites as an alternative structural material for enhanced strength, stiffness, damping, and multi-functionality. Modeling nano-composites involves modeling of nano-particles, and in the present case, we have used carbon nanotubes (CNT). We have shown that modeling CNT as a continuum one- or two-dimensional waveguide can simulate the atomistic wave behavior quite well. In addition, these materials can propagate waves in the order of Tera-Hz and at such frequencies, except for SFE, none of the other techniques work well. To the authors' best knowledge, there is no book in the archival literature that addresses these aspects. This has motivated us to include this topic in this book.

The aim of the book is twofold; first, it is to show that wavelet transform is indeed very useful in solving ordinary differential equations obtained by modeling the structure as a discrete system involving structural dynamics problems, and second, it is to use wavelet transform to solve those problems involving partial differential equations, which are hitherto not explored. The book also addresses in detail the complexity associated with wave propagation in both isotropic and anisotropic 1-D and 2-D structures due to boundary reflections, spurious dispersion, frequency, and time resolution of responses. The notable feature of the book is a chapter on the inverse problems, wherein the special advantage of using a wavelet-based spectral element is addressed in solving force identification and damage detection problems.

A step-by-step modular approach is adopted in writing this book. A number of numerical examples is presented not only to emphasize the efficiency of the method, but also to bring out the physics associated with each of these problems. The present method can be conveniently implemented under MATLAB environments and hence to aid the reader, a number of MATLAB scripts are given, which can be used to solve the problems addressed in the book. The material presented in this book can be used as a reference book for wave propagation in structure and, in addition, a separate graduate level course on wavelet-based analysis of structure can be developed. This book is written assuming the reader has only elementary knowledge of theory of elasticity, strength of materials, linear algebra, and methods of solving partial and ordinary differential equations.

The book was not possible without the help of many of our graduate and undergraduate students, who have contributed either directly or indirectly toward the development of this book. We would like to thank all of them.

S. Gopalakrishnan
Mira Mitra

Chapter 1

Introduction

Over the last two decades, wavelets are being effectively used for signal processing and solution of differential equations. There are several textbooks and articles in these areas. For more than a decade, wavelet transform has been implemented to solve and analyze problems associated with engineering mechanics. The use of wavelets in mechanics can be viewed from two perspectives, first, the analysis of mechanical responses for extraction of modal parameters, damage measures, de-noising, etc., and second, the solution of the differential equations governing the mechanical system. Though the wavelet transform has been proved to be very effective in all the above cases, there is no comprehensive text on the use of wavelet transform for solution of mechanics problems. The existing books are either on the mathematical aspects of wavelets, their use as approximation bases for solution of differential equations, or on the use of wavelets in signal processing. This book is an effort to explain wavelet transform and its applications from a mechanical engineering outlook. The book is written for graduate students and researchers working in the area of advanced structural dynamics and also numerical techniques for dynamic problems.

This book is about numerical solution of structural dynamics and wave propagation problems using wavelet transform methods. The book also has an introductory discussion on signal processing aspects of wavelet transform, including a comparison with other integral transforms like Laplace and Fourier transforms. Structural dynamics deal with lower frequencies in the magnitude of a few hundred hertz, or only the first few modes of vibration. In general, the steady state response is studied and, hence, the phase information is not very crucial. On the other hand, wave propagation results from high frequency excitations, in the order of kilohertz. Here, the transient response is studied and the phase information becomes essential. Apart from wavelet analysis of structural dynamics, this text mainly concentrates on wavelet-based spectral analysis of wave propagation. A new numerical scheme, *wavelet-based spectral finite element* method for modeling one, and two-dimensional structures is presented. This technique preserves the computational efficiency of spectral analysis, while possessing several advantages over Fourier transform-based spectral analysis particularly for capturing near field phenomena. In addition, the method can model relatively complex structures encountered in practice. The book includes chapters on wave propagation analysis of carbon nanotubes and their composites which is of much relevance for applications like sensors, oscillators, and other nano-scale devices. Finally, the application of wave propagation studies for structural health monitoring, force identification, and control of wave transmission is provided. To make the text comprehensive, a brief review of basic concepts of structural dynamics and wave propagation is included. To aid the readers to solve some problems, Matlab programs are included.

1.1 Solution of structural dynamics problem

The solution of structural dynamics problems can either be the determination of system parameters, mostly, natural frequencies and mode shapes, or simulations of the response of the system to external excitations that may be initial displacements, applied load, support motion, etc. For a discrete system, i.e., a multi-degree of freedom (MDOF) system, the governing equation is a set of ordinary differential equations (ODEs), coupled in general. Computational ease is achieved by decoupling these equations through linear transformation by modal matrix and is referred to as modal analysis. The modal analysis is in turn an eigenvalue analysis and several numerical techniques, for example, matrix iteration method, are prevalent for this purpose.

Governing equations for continuous systems like rod, beam, plates, etc., are partial differential equations with variation in spatial and temporal dimensions. Certain simple continuous systems can be solved analytically. Such analytical solutions, though exact, are restricted to very few simplified systems. There are many approximate techniques to solve more complicated structures. In general, such methods convert the continuous system into a discrete system. These approximate solution methods can be broadly divided into two groups. In the first group, the solution is assumed as a linear combination of known functions. In other words, the solution is a sum of known functions multiplied by unknown coefficients. The assumed solution is substituted in the governing differential equations for the continuous system and it reduces the partial differential equation (PDE) into a set of ODEs with the unknown coefficients as variables. The Rayleigh-Ritz and Galerkin methods fall in this group. In the other group of methods, the dynamics of the continuous system is expressed in terms of a large number of discrete points on the system. One of these methods is the finite element (FE) technique and probably is the most popular approximate method. In FE method, the continuum structure is divided into finite sub-domains called finite elements. Each element is connected to the surrounded element through nodes. The dynamics of the continuous system is expressed in terms of the displacements of the nodes, i.e., the continuous system is reduced to a MDOF system. The displacement within each of this sub-element (finite element) is approximated by some functions (mostly polynomial). The coefficients of these functions are obtained in terms of the displacements of the nodes. The total energy of the structure is then expressed as the sum of the energy of these elements. External forces are included using principle of virtual work. Lagrange's equations are then applied to the total energy and the equations of motion are obtained. The final equations exactly represent a MDOF system and the eigen parameters like natural frequencies and mode shapes are obtained through eigenvalue analysis. The popularity of FE method is due to its capability of modeling structures of high complexity and arbitrary geometry. There are several other methods like boundary element method, and mesh-less methods which have also been widely applied to solve structural dynamic systems.

This far, the numerical techniques available for modeling a dynamical systems required to obtain the system parameters have been discussed. In structural dynamics, a step after this is the simulation of the response of the system to external excitation. For MDOF system and the continuous system reduced to discrete system through approximate methods, the equations of motion are ODEs in time and mostly coupled. Though there are several standard methods for solution of these second order ODEs, in the realm of structural dynamics they are solved using direct integration scheme and mode superposition method. In direct integration schemes, namely, central difference method, Wilson θ method, Newmark's time integration, the time integration is performed without any transformation of the differential equations. Mode superposition method, in contrast, modifies the system of coupled governing differential equations into a set of decoupled ODEs which are much easier to solve. The decoupling is done through linear transformation and the transformation matrix is generally the eigenvector matrix of the MDOF system. The solution is obtained for the transformed variable referred to as generalized displacements and the displacements can be obtained through inverse transformation.

Another method of solution is the integral transform method. Here, the transformation, e.g., Laplace and Fourier transforms, is used in time. For certain problems, these transformations can be directly applied to the PDEs governing a continuous system. This reduces the PDEs by one dimensions and they can be solved through different approximate methods. In the other way, these transformations can be applied for time approximation in a MDOF system or a continuous system reduced to a MDOF system. In this book, Chapter 3, "Structural Dynamics: Introduction and Wavelet Transform," provides a brief review of the fundamentals of structural dynamics. In addition, the chapter describes the use of integral transforms, particularly wavelet transform, for obtaining the response of a MDOF system.

1.2   Solution of wave propagation problem

Wave propagation is a transient dynamic phenomenon resulting from short duration loading. Such transient loadings have high frequency content. The main difference between the structural dynamics and wave propagation in structures arises due to high frequency excitations in the later case. Structures very often experience such loadings in forms of impact and blast loadings like gust, bird hit, tool drops, etc. Apart from understanding the behavior of structures under such loading, wave propagation analysis is also important to gain knowledge about their high frequency characteristics, which have several applications. The applications include structural health monitoring using diagnostic waves and control of wave transmission for reduction of noise and vibration.

Though FE method is versatile and widely used to model complex structures of arbitrary geometry for structural dynamics problem, it is highly unsuited for wave propagation analysis. Higher frequency content of the loading in wave propagation problems requires very fine mesh with the element size comparable to the wavelengths, which are very small at higher frequencies. This results in large system size and huge computational cost. In addition to the fine mesh, to obtain system response, the mode superposition method or time integration schemes have to be implemented after FE modeling. Mode superposition method cannot be applied for wave propagation analysis. This is because for such problems the modal parameters have to be extracted over a wide range of frequency. This has to be done through eigenvalue analysis which is computationally very expensive. Alternatively there are several time integration schemes described earlier for solution of dynamics problems. These schemes can be used for simulation of wave response. In these methods, analysis is performed over a small time step, which is a fraction of the total time for which the response histories are required. For some time integration schemes, however, a constraint is placed on the time step, and this coupled with large system sizes makes the FE solution of wave propagation problems computationally prohibitive. Thus, in general, alternative numerical techniques are adopted for these problems and several such techniques are reported in literature. These include boundary element method (BEM) [7, 59], discontinuous Galerkin method [76, 40], meshless local Petrov-Galerkin (MLPG) method [4], wave finite element method [88], etc., to name a few.

Among these techniques, many methods are based on integral transform [37] which include Laplace transform, Fourier transform, and most recently wavelet transform. In these methods, first the governing equations are transformed to the frequency domain using the forward transform in time. Such transformation reduces the governing PDEs by one dimension to differential equations with only spatial variations. The solution of these transformed equations is much easier than the original PDEs and often has analytical solution. The main advantage of the system is the computational efficiency which is very less when compared to FE solution. These solutions in transformed frequency domain contain information of several frequency dependent wave properties essential for the analysis. The time domain solution is then obtained through inverse transform. The use of Laplace transform for solution of wave equation has been limited because of the difficulty in performing the inverse transform. On the other hand, the application of continuous Fourier transform (CFT) for such purpose has been reported in earlier work [89], but even here, the inverse transform required is difficult to obtain and, thus, these methods are suited only for far-field behavior like seismological studies. In structural wave propagation, the structures are finite, and hence these schemes are not adequate since, due to inherent problems in obtaining the transform, it cannot provide information about the reflection of waves on interaction between different boundaries and discontinuities. The forward and inverse discrete Fourier transform (DFT), however, can be numerically implemented. Fast Fourier transform (FFT) is the easy and fast algorithm for DFT. Spectral finite element (SFE) is one such method based on Fourier transform and initially proposed by Narayanan and Beskos [72] and popularized by Doyle and his co-researchers [28]. In brief, in SFE, the differential equations are reduced to ODEs using Fourier transform in time and FE procedure is followed in the transformed frequency domain. The solution gives the transformed displacements, which are converted to the displacements in time domain through inverse transform. Apart from being computationally efficient and capable of handling relatively complex structures, the SFE technique possesses an important advantage of simultaneous frequency domain analysis required to study the frequency dependent wave characteristics. The SFE technique like the other integral transform-based methods is effective in handling inverse problems like force identification and system identification.

From the mathematical explanation of wavelets and the literature available, it is clear that wavelets are potential candidates for spectral finite element formulation. There are several wavelets like Daubechies orthogonal wavelets, bi-orthogonal spline (B-spline) wavelets, interpolation wavelets, which have compactly supported bases with local supports and orthogonal properties. Thus they can be used to solve partial differential wave equations through integral transform because of the following advantages. Firstly, wavelet allows finite domain analysis and imposition of initial or boundary conditions, which are possible due to the local support of these basis functions. Secondly, as these bases are bounded both in time and frequency domains, frequency domain analysis can also be done apart from the time domain analysis. However, the resolution in frequency domain may be reduced when compared to Fourier transform-based SFE and this is the trade-off to obtain better resolution in time domain analysis. The results of these properties of wavelets are that the wavelet transform-based SFE remain computationally efficient yet will be able to model finite dimension waveguides for simulation of time response. This is otherwise not possible using the conventional Fourier transform-based

SFE. Apart from the use of wavelet methods for solution of structural dynamics problems, this book deals with the development of wavelet-based spectral finite element (WSFE) method and its applications.

1.3   Objective and outline of the book

The main objective of the book is to study the use and effectiveness of wavelet methods for solution of structural dynamics and wave propagation problems. The first part of the book studies the implementation of wavelet transform for simulation of the response of an MDOF system. The advantage of the wavelet transform method over the other common integral transforms is also explained. The next part of the book is about the development of a wavelet-based simulation scheme for wave propagation analysis. This novel technique is termed as "*wavelet-based Spectral Finite Element*" (WSFE) method. The WSFE scheme is formulated for 1-D and 2-D structural waveguides and is used for wave propagation both in time and frequency domains. With the advent of carbon nanotube (CNT) and their composite, referred as nano-composite, a lot of research is being directed towards understanding their mechanics. This is mainly because CNTs and nano-composites are foreseen to be used as an extraordinary material for different structures. The study of dynamics and wave propagation of CNTs and nano-composites is of very much relevance due to their various applications to sensors, oscillators, and many other devices. The book contains two chapters on vibration and wave propagation analysis of CNTs and nano-composites using WSFE. Finally, the applications of wave propagation to inverse problems of force and damage identification, and control of wave transmission are presented in the last chapter.

The book is organized into ten chapters including an introduction chapter. In Chapter 2, there is a brief introduction of integral transform methods which include Laplace, Fourier, and wavelet transforms. In wavelet transform, the emphasis is given to Daubechies compactly supported wavelets. The next chapter gives an introduction to the fundamental of structural dynamics and different methods to solve such problems. Here, the implementation of integral transform, in particular, wavelet transform, is discussed in the context of structural dynamics. In this regard, the imposition of initial conditions and differentiation of wavelet basis functions are provided. Chapter 4 is on the spectral analysis of wave motion. In this chapter, the characteristics of wave propagations like wavenumbers and wave speeds and also the methods to obtain them are discussed. The nature of wave transmission in elementary structural waveguides, for example, rod and beam, is studied here. The chapter also gives an introduction to Fourier transform-based SFE. Chapter 5 discusses the formulation of WSFE for 1-D waveguides including rod, Euler-Bernoulli beam, 2-D frame structures, and higher order composite beam. The developed method is used for simulation of time domain responses and is also validated with FE solution. The WSFE method is also compared with the corresponding Fourier transform-based technique to highlight the advantages of the former method in modeling finite waveguides. In Chapter 6, frequency domain analysis of wave propagation study with WSFE is explained. The extraction of frequency dependent wave characteristics is not very straightforward with the wavelet transform, in contrast to Fourier transform. In WSFE, there exist a constraint on the time sampling rate depending on the frequency content of the excitation. This chapter explains this constraint which needs to be satisfied to avoid spurious dispersion in the time domain solution. Chapter 7 is on the development of WSFE for 2-D waveguides, which need wavelet transform both in time and one spatial direction. The 2-D waveguides studied are plate and axisymmetric cylinder. Even here, WSFE is used for time and frequency domain analysis. In Chapter 8, the vibration and wave propagation is studied for CNTs both single-walled and multi-walled. The solution is obtained using WSFE and the CNTs are modeled using both beam and shell theory. According to the available literature, such continuum models of CNTs give appreciable results when compared to the corresponding experimental results existing. Chapter 9 provides similar analysis, but for nano-composite structures. Nano-composites consist of a matrix material, which can be polymer, metal, or ceramic with CNT embedded in it. Finally, the last chapter provides three important examples of applications of wave propagation. First, the problem of force identification from the measured wave responses in 1-D waveguides. Second, the modeling of de-lamination in composite beam and detection of the damage inversely from simulated wave responses acting as surrogate experimental results. A wavelet-based damage detection scheme is employed for this purpose. Third is the passive control of wave transmission in 1-D waveguides. The control is achieved as the reduction in the transmission of elastic power resulting from wave motion from the source to the other parts of the structure. A nano-composite insert is used, which because of high stiffness and low density helps in preventing the flow of power from one region to another.

Chapter 2

Integral Transform Methods

In this chapter, a brief review of integral transform methods is presented. The methods described are Laplace transform, Fourier transform, and wavelet transform. Though several textbooks are available on these methods, this chapter is intended to give a more engineering view to this subject. It also aims to make the readers acquainted with the different terms and their definitions which are used in the later part of the book.

In a broad sense, the idea of integral transform is to represent a function $F(t)$, in terms of known functions. The problem lies with how efficiently the function $F(t)$ can be expressed in terms of *simple* or *elementary* functions. The choice of this family of known functions depends on several issues like the ease of the transformation, purpose of the analysis, etc. The *transform* of the function $F(t)$ can be written as

$$\check{F}(\alpha) = \int_a^b F(t)E(\alpha, t)dt \qquad (2.1)$$

$E(\alpha, t)$ is the known function of α and t and is referred as the *kernel* function. For finite a and b, the *transform* $\check{F}(\alpha)$ is a *finite transform*.

The transformation is either needed for better analysis of the function $F(t)$ in a *transform domain* with the variable α in this case. Otherwise, the transformation is used to solve boundary-value problems posed by a differential equation. The *integral transform* reduces the number of dimensions of differential equations. These reduced equations are in most cases easier to solve. The use of integral transform for solution of differential equations in the context of structural dynamics is presented in the next few chapters.

2.1 Laplace transform

Let us consider a time signal $F(t)$, which is defined for all $t > 0$. The *Laplace transform* of $F(t)$ is obtained as

$$\bar{F}(s) = \mathscr{L}F(t) = \int_0^\infty e^{-st}F(t)dt \qquad (2.2)$$

Here, s is a subsidiary variable and in general a complex quantity. The original time signal $F(t)$ can be obtained through an *inverse transform* of $\bar{F}(s)$ and is written as

$$F(t) = \mathscr{L}^{-1}\bar{F}(s) \tag{2.3}$$

The Laplace transform of $F(t)$ exists for all $s > \gamma$, only for $F(t)$ being *piecewise continuous* on any finite interval in the range $t > 0$ and

$$|F(t)| < Ce^{\gamma t} \tag{2.4}$$

for some constants C and γ. Almost all physical signals satisfy the above constraint.

The Laplace transformation is a linear operation. For any functions $F(t)$ and $G(t)$ whose Laplace transform exists and for any constants a and b,

$$\mathscr{L}\{aF(t) + bG(t)\} = a\mathscr{L}F(t) + b\mathscr{L}G(t) \tag{2.5}$$

To show an example of Laplace transform, let us consider $F(t) = 1$ for $t > 0$. The Laplace transform is

$$\check{F}(s) = \mathscr{L}F(t) = \mathscr{L}(1) = \int_0^\infty e^{-st}dt = -\frac{1}{s}e^{-st}\bigg|_0^\infty \tag{2.6}$$

when $s > 0$,

$$\mathscr{L}(1) = \frac{1}{s} \tag{2.7}$$

Next, let us consider an example of rectangular pulse in time, defined as,

$$F(t) = \begin{cases} F_0 & t_0 \le t \le t_0 + d \\ 0 & \text{otherwise} \end{cases} \tag{2.8}$$

The Laplace transform of $F(t)$ is obtained as

$$\check{F}(s) = \int_{t_0}^{t_0+d} F_0 e^{-st}dt = -\frac{F_0}{s}e^{-st}\bigg|_{t_0}^{t_0+d} = \frac{F_0 e^{-st_0}}{s}\left(1 - e^{-sd}\right) \tag{2.9}$$

The Laplace transform can be conveniently used for solutions of linear ODEs with constant coefficients. These kinds of equations are often encountered in structural dynamics. This transform, however, can be performed only analytically and there is no numerical implementation. This restricts the use of Laplace transform for analysis of problems with higher complexities, which need to be solved numerically.

2.2   Fourier transform

Probably the most used integral transform to date is the Fourier transform. It is well known that Fourier transform decomposes a signal into its sinusoidal components. One of the main advantages of using Fourier transform for structural dynamics and wave propagation problems is that several important characteristics of the system can be directly obtained from the transformed frequency domain. In addition, Fourier transform in principle can achieve high accuracy in differentiation and thus can be used for solution of differential equations. The Fourier transform can be implemented analytically, semi-analytically and numerically in the form of *continuous Fourier transform* (CFT), *Fourier series* (FS), and *discrete Fourier transform* (DFT) respectively. In this section, apart from these three Fourier transform methods, an introduction is also given for *short term Fourier transform* (STFT), which, in brief, performs the Fourier transform of truncated or windowed portions of the signal.

2.2.1 Continuous Fourier transform

The forward and inverse CFTs of the time signal $F(t)$ can be written as

$$\widehat{F}(\omega) = \int_{-\infty}^{\infty} F(t)e^{-\jmath\omega t}dt \qquad F(t) = \frac{1}{2\pi}\int_{-\infty}^{\infty} \widehat{F}(\omega)e^{\jmath\omega t}d\omega \qquad (2.10)$$

where $\widehat{F}(\omega)$ is referred as the CFT of $F(t)$, ω is the angular frequency and $\jmath = \sqrt{-1}$. $\widehat{F}(\omega)$ is in general complex and a plot of the amplitude of this function against frequency gives the spectral density of the time signal $F(t)$. As an example let us consider the rectangular pulse given by Equation 2.8. The CFT is obtained using Equation 2.10,

$$\widehat{F}(\omega) = \int_{t_0}^{t_0+d} F_0 e^{-\jmath\omega t}dt = F_0 d\left(\frac{\sin\omega d/2}{\omega d/2}\right) e^{-\jmath\omega(t_0+d/2)} \qquad (2.11)$$

Following are some of the important properties of CFT

- **Linearity:** The CFT is a linear transformation, i.e., for two functions $F(t)$ and $G(t)$, and constants a and b, the Fourier transform of the function $aF(t)+bG(t)$ can be obtained as $a\widehat{F}(\omega)+b\widehat{G}(\omega)$. This can also be written as $aF(t) + bG(t) \Leftrightarrow a\widehat{F}(\omega) + b\widehat{G}(\omega)$ where the symbol $\Leftrightarrow$ denotes Fourier transformation hereafter.

- **Scaling:** If a time signal is multiplied by a non-zero factor k to become $F(kt)$, the CFT is given as $F(kt) \Leftrightarrow \frac{1}{|k|}\widehat{F}(\omega/k)$. This implies that compression in time domain results in dilation in frequency domain. The amplitude however decreases to keep the energy constant.

- **Shifting:** A shift in the time signal by t_s is manifested as a phase change in the transformed frequency domain obtained through CFT. The transform pair can be written as $F(t - t_s) \Leftrightarrow \widehat{F}(\omega)e^{-\jmath\omega t_s}$

- **Symmetric property of CFT:** The time signal $F(t)$ can always be written as superposition of a symmetric (even) and anti-symmetric (odd) function, $F_e(t)$ and $F_o(t)$ respectively, where

$$F_e(t) = \frac{1}{2} \left[F(t) + F(-t) \right], \qquad F_o(t) = \frac{1}{2} \left[F(t) - F(-t) \right]$$

CFT being linear and having the scaling property,

$$F_e(t) \Leftrightarrow \frac{1}{2} \left[\widehat{F}(\omega) + \widehat{F}(-\omega) \right] = Real \left[\widehat{F}(\omega) \right] = \widehat{F}_R(\omega) \qquad (2.12)$$

$$F_o(t) \Leftrightarrow \frac{1}{2} \left[\widehat{F}(\omega) - \widehat{F}(-\omega) \right] = \jmath Imag \left[\widehat{F}(\omega) \right] = \jmath \widehat{F}_I(\omega) \qquad (2.13)$$

Thus, combining Equations 2.12 and 2.13,

$$F(t) \Leftrightarrow \widehat{F}_R(\omega) + \jmath \widehat{F}_I(\omega) \qquad (2.14)$$

Substituting Equation 2.14 into the first part of Equation 2.10 and expanding $e^{-\jmath \omega t}$ in terms of sine and cosine functions, we get

$$\widehat{F}_R(\omega) = \frac{1}{2\pi} \int_{-\infty}^{\infty} F(t) \cos(\omega t) dt \qquad \widehat{F}_I(\omega) = \frac{1}{2\pi} \int_{-\infty}^{\infty} F(t) \sin(\omega t) dt$$

$$(2.15)$$

The first integral is an even function and the second is an odd function, i.e., $\widehat{F}_R(\omega) = \widehat{F}_R(-\omega)$ and $\widehat{F}_I(\omega) = -\widehat{F}_I(-\omega)$. Thus, $\widehat{F}(\omega) = \widehat{F}_R(\omega) + \jmath \widehat{F}_I(\omega)$ and $\widehat{F}(-\omega) = \widehat{F}_R(-\omega) + \jmath \widehat{F}_I(-\omega) = \widehat{F}_R(\omega) - \jmath \widehat{F}_I(\omega) = \widehat{F}^*(\omega)$. Now, $\widehat{F}(-\omega)$ is the CFT on the left of the origin ($\omega = 0$ here) and is the complex conjugate of the CFT $\widehat{F}(\omega)$ on the right of the origin. This origin point is called the Nyquist frequency and is of much importance as the Fourier transform is required to evaluate only for half of the total frequency range. The CFT on the other side of the Nyquist frequency is obtained as the conjugate of the other half for any real function as the time signal.

- **Convolution:** A very useful and interesting property of CFT is that the convolution of two signals in time domain is their product in transformed frequency domain and vice versa. The CFT of two time signals $F_1(t)$ and $F_2(t)$ is obtained as

$$\widehat{F}_{12}(\omega) = \int_{-\infty}^{\infty} F_1(t) F_2(t) e^{-\jmath \omega t} dt \qquad (2.16)$$

Substituting Equation 2.10 in the above equation for both these functions, we get

$$\widehat{F}_{12}(\omega) = \int_{\infty}^{\infty} \widehat{F}_1(\bar{\omega}) \int_{-\infty}^{\infty} F_2(t) e^{-\jmath(\omega - \bar{\omega})t} dt d\bar{\omega} = \int_{-\infty}^{\infty} \widehat{F}_1(\bar{\omega}) \widehat{F}_2(\omega - \bar{\omega}) d\bar{\omega}$$

$$(2.17)$$

This can be written as

$$F_1(t)F_2(t) \Leftrightarrow \int_{-\infty}^{\infty} \widehat{F}_1(\omega)\widehat{F}_2(\omega - \bar{\omega})d\bar{\omega} \qquad (2.18)$$

and, conversely, $\widehat{F}_1(\omega)\widehat{F}_2(\omega)$.

2.2.2 Discrete Fourier transform

The continuous Fourier transform can only be applied to analytical functions, for example, signals which are given as continuous functions of time. Thus, it cannot be used for numerical analysis. This is a serious limitation as majority of the present day problems are required to be solved *numerically*. The same example of the signal can be considered where it is obtained as numerical data captured at certain time interval as it happens in most of the real life situations. This necessitates a numerical representation of Fourier transform and is termed as discrete Fourier transform (DFT). There is however an intermediate form, the Fourier series (FS), where the inverse transform is written in form of series as

$$F(t) = \frac{1}{2}a_0 + \sum_{n=1}^{\infty}\left[a_n\cos\left(2\pi n\frac{t}{T}\right) + b_n\sin\left(2\pi n\frac{t}{T}\right)\right] \text{ T is the period of } F(t)$$

$$(2.19)$$

It should be noted that the numerical representation of Fourier transform in FS and also in DFT requires a periodicity assumption. The signal is assumed to have a time period T after which it repeats itself. The FS coefficients a_n and b_n are obtained from forward transform which is written in integral form,

$$a_n = \frac{2}{T}\int_0^T F(t)\cos\left(2\pi n\frac{t}{T}\right)dt \quad b_n = \frac{2}{T}\int_0^T F(t)\sin\left(2\pi n\frac{t}{T}\right)dt$$

$$n = 0,\ 1,\ 2,\ldots \qquad (2.20)$$

Using the symmetric and anti-symmetric properties of a_n and b_n respectively, Equation 2.19 can be rewritten in following exponential form,

$$F(t) = \frac{1}{2}\sum_{-\infty}^{\infty}(a_n - \imath b_n)e^{-\imath\omega_n t} = \sum_{-\infty}^{\infty}\widehat{F}_n e^{\imath\omega_n t} \quad \omega_n = \frac{2\pi n}{T} \qquad (2.21)$$

and

$$\widehat{F}_n = \frac{1}{2}(a_n - \imath b_n) = \frac{1}{T}\int_0^T F(t)e^{-\imath\omega_n t}dt \quad n = 0,\ \pm 1,\ \pm 2\ldots \qquad (2.22)$$

The main aim of DFT is to replace the integral form of the forward Fourier transform given by Equation 2.22 by a summation for numerical implementation. This can be done in the following way. Let us consider the time signal

$F(t)$ is divided into M equal width rectangles with height F_m which is the value of $F(t)$ at any time instant t_m, $m = 0,\ 1,\ldots,M-1$. The width is the time interval $\Delta t = \frac{T}{M}$. Now knowing that the CFT of a rectangle is a *sinc* function, with the rectangular approximation of the signal, the integral given by the Equation 2.22 can be written as the summation of M *sinc* functions of pulse width ΔT as follows

$$\widehat{F}_n = \Delta T \left[\frac{\sin(\omega_n \Delta T/2)}{(\omega_n \Delta T/2)}\right] \sum_{m=0}^{M} F_m e^{-\imath\omega_n t_m} \tag{2.23}$$

For fine discretization, ΔT is very small which make the value of the *sinc* function given in Equation 2.23 nearly equal to unity. Hence the forward and inverse DFT can be written as

$$\widehat{F}_n = \widehat{F}(\omega_n) = \Delta T \sum_{m=0}^{N-1} F_m e^{-\imath\omega_n t_m} = \Delta T \sum_{m=0}^{N-1} F_m e^{-\imath 2\pi nm/N}$$

$$F_m = F(t_m) = \frac{1}{T} \sum_{n=0}^{N-1} \widehat{F}_n e^{\imath\omega_n t_m} = \frac{1}{T} \sum_{n=0}^{N-1} \widehat{F}_n e^{\imath 2\pi nm/N} \tag{2.24}$$

Here, both n and m range from 0 to $N-1$.

Similar to CFT, the DFT of a real function, for example a time signal, is also symmetric about the Nyquist frequency. The DFT coefficients on one side of the Nyquist frequency is the complex conjugate of the coefficients on the other side of it. Thus, even here, for a real function, the DFT is required to be evaluated only for half of the frequency range, i.e., up to the Nyquist frequency given as $f_{nyq} = \frac{1}{2\Delta T}$.

In this subsection and following subsections on *windowed Fourier transform* and *discrete wavelet transform*, a comparative study of the three methods in the context of spectral analysis of time signals is presented [74]. All these transforms can provide spectral analysis of a signal but vary from each other with respect to the time resolution. Fourier transform decomposes a signal into its frequency components and the coefficients provide the averaged spectral density independent of time. Hence, the Fourier transform of two signals with different time histories can show the same spectral density [74]. Let us consider a $N = 16$ points time signal shown in Figure 2.1(a) with sampling time interval $\Delta T = 0.004$ s. Figure 2.1(b) shows the corresponding DFT coefficients $|\widehat{F}_n|$. According to what we discussed before, the moduli of the DFT coefficients are symmetric about the Nyquist frequency. Modulus of each of the DFT coefficients $|\widehat{F}_n|$ for $n = 0,\ 1,\ldots,N-1$ denotes the spectral density of the signal averaged over the time period T at the circular frequency $\omega_n = \frac{2\pi n}{N\Delta T}$.

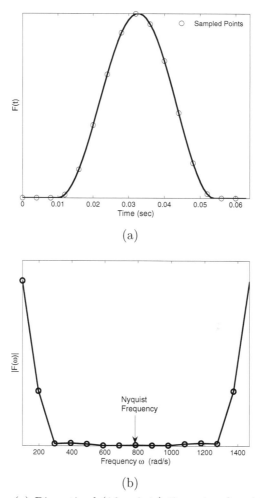

FIGURE 2.1: (a) Discretized (16 points) time signal and (b) DFT coefficients of the signal

2.2.3 Short term Fourier transform

The short term Fourier transform (STFT) replaces the sinusoidal approximation of a signal by the product of the sinusoid and a window localized in time. Thus, it has two arguments, time and frequency. The STFT is defined as

$$\widehat{F}^{SF}(t_0,\omega) = \int_{-\infty}^{\infty} F(t)G(t-t_0)e^{-\imath\omega t}dt \qquad (2.25)$$

Here, $G(t)$ is the windowing function centered at time t_0. This window captures a short length of the data centered at $t = t_0$ and only this part of the signal contributes to $\widehat{F}^{SF}(t_0,\omega)$. The window can be moved along the time axis and the calculation is repeated. In this way, a time-frequency distribution of the signal can be achieved which can circumvent the drawback of Fourier transform which provide the frequency information averaged over the entire length of time.

In frequency domain, the STFT or the Fourier transform of the truncated signal can be obtained as the convolution of the Fourier transform of the original signal $F(t)$ and the Fourier transform of the windowing function $G(t)$ and can be written as

$$\widehat{F}^{SF}(t_0,\omega) = \frac{1}{2\pi}\widehat{F}(\omega) * \widehat{G}(\omega) \qquad (2.26)$$

Thus, even in frequency domain the STFT results in selecting a frequency window defined by $\widehat{G}(\omega)$ from the entire frequency content of the signal given by $\widehat{F}(\omega)$. Thus the properties of STFT are determined by the window, or rather its Fourier transform. In order to preserve the properties of the original signal to the most, the Fourier transform of $G(t)$ should be real, even, and the energy should be concentrated around the Nyquist frequency. The most rudimentary form is the rectangular window. A rectangular window over the interval $[0\ T_w]$ is defined as,

$$G(t) = \begin{cases} 1 \text{ for } & 0 \le t \le T_w \\ 0 & \text{otherwise} \end{cases} \qquad (2.27)$$

A rectangular window cut the signal into non-overlapping slices of length T_w. Performing STFT using rectangular window is equivalent to doing Fourier transform on each slice of the signal separately. Such abrupt cut-off of the signal, however, results in some unwanted features. Thus, more smooth windows are used and some of the classical windowing functions normalized as $G(0) = 1$ are presented in Table 2.1.

For a window with width T_w, the frequency bandwidth is approximately $\frac{1}{T}$. Thus, using a shorter window means that the bandwidth is wider. In other words, it is not possible to obtain a higher degree of resolution both in time and frequency. A feature of STFT is that it has same time resolution at all the frequencies. To obtain finer time resolution in higher frequency will require to

TABLE 2.1: Different windowing functions, $G(t)$ used for WFT

	$G(t)$
Hamming	$0.54 + 0.46\cos(2\pi t)$
Gaussian	e^{-18t^2}
Hanning	$\cos^2(\pi t)$
Blackman	$0.42 + 0.5\cos(2\pi t) + 0.08\cos(4\pi t)$

keep the same resolution at the lower frequencies which is redundant. These drawbacks of Fourier transform and also short term Fourier transform are circumvented by wavelet transform explained in the next section.

2.3 Wavelet transform

The word *wavelet* has been derived from the French word *ondelette* meaning "small wave" and coined by Morlet and Grossmann [70, 69, 38] in early 1980s. The development of wavelet like functions, notably, Haar wavelets and Littlewood-Paley wavelets, however, date backs to the early part of 20th century. Since then from around 1975, a lot of researchers were involved in development of wavelets. Some notable contributors include Morlet and Grossmann [38] for formulation of continuous wavelet transform (CWT), Stromberg [93] for early works on discrete wavelet transform (DWT), Meyer [61] and Mallat [58] for multi-resolution analysis using wavelet transform, and Daubechies [20] for proposal of orthogonal compactly supported wavelets. Thereafter, a lot of work has been done both on development and application of wavelet analysis on a wide variety of problems like signal and image processing, data condensation, solution of differential equations. In this book, we will primarily concentrate on use of wavelet transform for signal processing in the realm of structural dynamics and also the solution of governing equations of motion of structural systems. In the following subsections, we provide a generalized introduction to wavelets and then focus on the details of Daubechies orthogonal compactly supported wavelets.

According to Daubechies [20], wavelet transform can be defined as, "*The wavelet transform is a tool that cuts up data, functions or operators into different frequency components, and then studies each component with a resolution matched to its scale.*" For example, in analysis of time signal for structural dynamics, the wavelet transform will decompose the signal into its frequency components and for each of these frequency components, the time histories are preserved, but with a certain time resolution depending on the frequency. Hence, unlike STFT given by Equation 2.25, the forward continuous wavelet transform (CWT) of the $F(t)$ is written as

$$F^W(a, b) = \int_{-\infty}^{\infty} F(t)\psi\left(\frac{t-b}{a}\right) \tag{2.28}$$

Here, $\psi(t)$ is the wavelet basis function and can be thought of as a windowing function similar to $G(t)$ for STFT. b defines the position in time and a defines the width of $\psi(t)$. These parameters modify the basis function $\psi(t)$ by scaling and shifting it.

2.3.1   Multi-resolution analysis with wavelets

Prior to studying the use of wavelets for multi-resolution analysis (MRA), we should understand its mathematical basis. For multi-resolution representation of a function in $L^2(\mathbb{R})$, we need to obtain a sequence of closed subspaces V_j for $j \in \mathbb{Z}$ with the following properties,

1. $V_j \subset V_{j+1} \; \forall \; \mathbb{Z}$, i.e.,

$$\{0\} \cdots \subset V_{-1} \subset V_0 \subset V_1 \subset V_2 \cdots \subset L^2(\mathbb{R})$$

2. $\bigcup_{j \in \mathbb{Z}} V_j$ is dense in $L^2(\mathbb{R})$

3. $\bigcap_{j \in \mathbb{Z}} V_j = \{0\}$

4. The subspaces are related by a scaling relation,

$$F(t) \in V_j \Leftrightarrow F(2t) \in V_{j+1} \quad \forall \; j \in \mathbb{Z}$$

    This can also be written as

$$F(t) \in V_j \Leftrightarrow F(2^k t) \in V_{j+k} \;\; \text{and} \;\; F(t) \in V_j \Leftrightarrow F(2^{-j}t) \in V_0 \;\; \forall \; j \in \mathbb{Z}$$

    Thus, the problem of finding the embedded subspaces V_j essentially reduces to the problem of obtaining V_0.

5. Each subspace is spanned by integer translates of a single function,

$$F(t) \in V_j \Leftrightarrow F(t+1) \in V_j \;\; \forall j \in \mathbb{Z}$$

From the above properties of MRA it can be concluded that we need to find a scaling function $\varphi(t) \in V_0$ such that its integer translates $\{\varphi(t-k), k \in \mathbb{Z}\}$ are the Riesz bases for the space V_0. Now, $\varphi(2t-k)$ will form a basis for the space V_1. Thus,

$$V_0 = \overline{\text{span}\{\varphi(t-k), k \in \mathbb{Z}\}}$$
$$V_1 = \overline{\text{span}\{\varphi(2t-k), k \in \mathbb{Z}\}}$$

Since $V_0 \in V_1$, the basis functions of space V_0 can be expressed in terms of the basis functions in V_1 as

$$\varphi(t) = \sum_{-\infty}^{\infty} a_k \varphi(2t - k) \qquad (2.29)$$

Equation 2.29 is referred to as dilation or scaling relation and a_k, $k \in \mathbb{Z}$ are referred to as filter coefficients. The above equation can be solved to derive the scaling function $\varphi(t)$ which forms the bases for space V_0. The basis function for V_j can thus be defined as

$$\varphi_{j,k} = 2^{j/2} \varphi(2^j t - k) \qquad (2.30)$$

Here, j and k are the dilation and translation indices. In analysis of time signals, j corresponds to the frequency and k to the time.

Let us denote the approximation of the function F(t) by the scaling functions $\varphi_{j,k}(t)$ as $P_j F$. In other words, it is the projection of $F(t)$ into the subspace V_j,

$$P_j F = \sum_{-\infty}^{\infty} c_{j,k} \varphi_{j,k}(t) \qquad (2.31)$$

Here, $c_{j,k}$ are the *approximation coefficients* and as $j \to \infty$, $P_j F \to F$.

The next step is to obtain a closure subspace W_j, $j \in \mathbb{Z}$ for the subspaces V_j and its orthogonal complement such that

$$V_{j+1} = V_j \oplus W_j \quad \text{and} \quad V_j \perp W_j \qquad (2.32)$$

Here, $\oplus$ denotes the direct sum. The subspaces W_j are orthogonal and also

$$\oplus_{j \in \mathbb{Z}} W_j = L^2(\mathbb{R}) \qquad (2.33)$$

The wavelet function $\psi(t)$ is defined such that it translates $\psi(t - k)$, $k \in \mathbb{Z}$ are the Riesz bases for W_0. Thus,

$$\psi_{j,k} = 2^{j/2} \psi(2^j t - k) \qquad (2.34)$$

form the Riesz basis for the subspace W_j. Similar to scaling function $\varphi(t)$ for subspace V_0, the wavelet function $\psi(t)$ for the subspace W_0 can be written as a linear combination of the basis functions for V_1 as $W_0 \subset V_1$,

$$\psi(t) = \sum_{-\infty}^{\infty} b_k \varphi(2t - k) \qquad (2.35)$$

Let us consider $Q_j F$ as the approximation of $F(t)$ using wavelet functions $\psi_{j,k}(t)$ and it is the projection of $F(t)$ on the subspace W_j. This can be written as

$$Q_j F = \sum_{-\infty}^{\infty} d_{j,k} \psi_{j,k}(t) \qquad (2.36)$$

Here, $d_{j,k}$ are referred as *detail coefficients*. Now, using Equation 2.32, $P_{j+1}F$ can be written as

$$P_{j+1}F = P_jF + Q_jF \tag{2.37}$$

Thus, the approximation of $F(t)$ at the higher or refined scale is obtained from the approximations at the lower scale with lower resolution. This forms the basis of multi-resolution analysis using wavelets.

2.3.2 Daubechies compactly supported wavelets

It can be summarized from what has been discussed in the previous subsection that wavelets $\psi_{j,k}$ form the basis functions for $L^2(\mathbb{R})$ and any function in $L^2(\mathbb{R})$ can be represented using these bases. Several wavelet functions, for example *Morlet wavelets*, *Shanon wavelets*, *Meyer wavelets*, *Mexican hat wavelets*, have been proposed by the researchers. The choice of the wavelet, however, depends on the nature of analysis to be performed. This book deals with the use of wavelet methods for both analysis of time signal in the context of structural dynamics and also for the solution of the wave equations to study elastic wave propagation in structural waveguides. The solution of such partial differential wave equations requires wavelet bases with localized supports for imposition of the boundary conditions. Daubechies [19, 20] proposed orthogonal compactly supported wavelets referred as Daubechies wavelets. There are other compactly supported wavelets like *bi-orthogonal spline (B-spline) wavelets* [16, 17], and *interpolation wavelets* [6, 22].

2.3.2.1 Construction of Daubechies compactly supported wavelets

This section is focussed on the properties and construction of Daubechies compactly supported wavelets. The first step in the derivation of these wavelets is to obtain the scaling functions $\varphi(t)$ from the scaling or dilation equation given by Equation 2.28. The filter coefficients a_k determine the nature of the wavelet function and for Daubechies compactly supported wavelets only a finite number of filter coefficients are non-zero. $\psi(t)$ is again obtained from $\varphi(t)$ using Equation 2.35 and for Daubechies wavelets it can be written as

$$\psi(t) = \sum_{-\infty}^{\infty} (-1)^k a_{1-k} \varphi(2t - k) \tag{2.38}$$

a_k and $(-1)^k a_k$ form the *quadrature mirror filters*. The above equation satisfies the orthogonal condition of scaling and wavelet functions required by the Equation 2.32. The filter coefficients are obtained by imposing the following constraints on the scaling functions,

1. For uniqueness, normalization is done by considering the area under the scaling function to be unity,

$$\int_{-\infty}^{\infty} \varphi(t)dt = 1 \tag{2.39}$$

The above equation leads to the following condition on the filter coefficients,

$$\sum_{-\infty}^{\infty} a_k = 2 \tag{2.40}$$

2. For Daubechies wavelets, the integer translates of scaling functions are orthogonal, i.e.,

$$\int_{-\infty}^{\infty} \varphi(t)\varphi(t+l)dt = \delta_{0,l}, \quad l \in \mathbb{Z} \tag{2.41}$$

where

$$\delta_{0,l} = \begin{cases} 1, & l = 0 \\ 0, & \text{otherwise} \end{cases}$$

This gives a condition on the filter coefficients as

$$\sum_{k=-\infty}^{\infty} a_k a_{k+2l} = 2\delta_{0,l}, \quad l \in \mathbb{Z} \tag{2.42}$$

3. The conditions on the filter coefficients given by Equations 2.40 and 2.42 do not give an unique set of filter coefficients. For an N coefficients system, the Equations 2.40 and 2.42 provide only $N/2 + 1$ equations for the filter coefficients. For the remaining $N/2 - 1$ equations required to obtain an unique set of filter coefficients we need some other conditions to be imposed on the wavelet functions. For Daubechies wavelets, the conditions are imposed by considering that the scaling functions should be able to exactly represent polynomials of order M and $M = N/2$. Let us consider a polynomial of order M as

$$f(t) = a_0 + a_1 t + a_2 t^2 + \cdots + a_{M-1} t^{M-1} \tag{2.43}$$

The above polynomial should be exactly represented by an expansion similar to that given by Equation 2.30 for $j = 0$ and can be written as

$$f(t) = \sum_{k=-\infty}^{\infty} c_k \varphi(t - k) \tag{2.44}$$

Since $\psi(t)$ are orthogonal to the translates of $\varphi(t)$, taking inner product of Equation 2.44 with $\psi(t)$ gives

$$\langle f(t), \psi(t) \rangle = \sum_{k=-\infty}^{\infty} c_k \langle \varphi(t-k), \psi(t) \rangle \equiv 0 \tag{2.45}$$

Thus, substituting Equation 2.43 in the above Equation 2.45, we get

$$a_0 \int_{-\infty}^{\infty} \psi(t)dt + a_1 \int_{-\infty}^{\infty} \psi(t)tdt + \cdots + a_{M-1} \int_{-\infty}^{\infty} \psi(t)t^M dt \equiv 0 \quad (2.46)$$

The above identity is valid for all a_j for $j = 0, 1, 2, \cdots M - 1$. Choosing $a_l = 1$ and all other $a_j = 0$ gives

$$\int_{-\infty}^{\infty} \psi(t)t^l dt \quad l = 0, 1, 2 \ldots M - 1 \qquad (2.47)$$

This implies that the first M moments of the wavelet function should be zero. Equation 2.47 can be written in terms of the filter coefficients after some calculations as

$$\sum_{k=-\infty}^{\infty} (-1)^k a_k k^l = 0 \quad l = 0, 1, 2, \ldots, M - 1 \qquad (2.48)$$

$N = 2M$ determines the order of the Daubechies wavelet and is referred as D4, D6, D8 and thereafter for $N = 4, 6, 8$ respectively.

As said before, the scaling functions are obtained by solving recursively the dilation Equation 2.29 which can be expanded for DN as,

$$\varphi(t) = a_0 \varphi(2t) + a_1 \varphi(2t - 1) + \cdots + a_{N-1} \varphi(2t - N + 1) \qquad (2.49)$$

Using the compactness criteria of the Daubechies scaling functions between 0 to $N - 1$ where N is the order of the DN Daubechies scaling function, the relation given in Equation 2.49 can be written as the following equations,

$$\varphi(0) = a_0 \varphi(0)$$
$$\varphi(1) = a_0 \varphi(2) + a_1 \varphi(1) + a_2 \varphi(0)$$
$$\varphi(2) = a_0 \varphi(4) + a_1 \varphi(3) + a_2 \varphi(2) + a_3 \varphi(1) + a_4 \varphi(0)$$
$$\vdots$$
$$\varphi(N - 2) = a_{N-3}\varphi(N - 1) + a_{N-2}\varphi(N - 2) + a_{N-1}\varphi(N - 3)$$
$$\varphi(N - 1) = a_{N-1}\varphi(N - 1)$$

This can also be written as matrix form,

$$\begin{bmatrix} a_0 & 0 & 0 & \cdots & 0 & 0 & 0 \\ a_2 & a_1 & a_0 & \cdots & 0 & 0 & 0 \\ a_4 & a_3 & a_2 & \cdots & 0 & 0 & 0 \\ \cdots & \cdots & \cdots & \cdots & \cdots & \cdots & \cdots \\ 0 & 0 & 0 & \cdots & a_{N-3} & a_{N-4} & a_{N-5} \\ 0 & 0 & 0 & \cdots & a_{N-1} & a_{N-2} & a_{N-3} \\ 0 & 0 & 0 & \cdots & 0 & 0 & a_{N-1} \end{bmatrix} \begin{bmatrix} \varphi(0) \\ \varphi(1) \\ \varphi(2) \\ \cdots \\ \varphi(N-3) \\ \varphi(N-2) \\ \varphi(N-1) \end{bmatrix} = \begin{bmatrix} \varphi(0) \\ \varphi(1) \\ \varphi(2) \\ \cdots \\ \varphi(N-3) \\ \varphi(N-2) \\ \varphi(N-1) \end{bmatrix} \qquad (2.50)$$

TABLE 2.2: Filter coefficients a_k for Daubechies scaling function with $N = 4$, 6, and 12

k	D4	D6	D12
0	0.68301270189244	0.47046720778540	0.07887121600143
1	1.18301270189174	1.14111691583462	0.34975190703757
2	0.31698729810756	0.65036500052742	0.53113187994121
3	-0.18301270189174	-0.19093441556852	0.22291566146505
4		-0.12083220831070	-0.15999329944587
5		0.04981749973178	-0.09175903203003
6			0.06894404648720
7			0.01946160485396
8			-0.02233187416548
9			0.00039162557603
10			0.00337803118151
11			-0.00076176690258

or

$$\mathbf{A}\varphi = \varphi$$

Thus, Equation 2.50 possesses an eigenvalue problem and can be solved to obtain φ as the eigenvectors. The matrix $\mathbf{A}$ is known as the filter coefficients a_k and can be solved from Equations 2.40, 2.42, and 2.48. The filter coefficients for D4, D6, and D8 are given in Table 2.2. These filter coefficients can be obtained using the MATLAB wavelet toolbox function *dbwavf*. The MATLAB program 2.1 shows the use of *dbwavf*.

Once the values of $\varphi(t)$ are known at the integer values of t between 0 to $N - 1$, the values at the points in between the integers can be obtained from the Equation 2.29 modified as

$$\varphi\left(\frac{t}{2}\right) = \sum_{k=-\infty}^{\infty} a_k\varphi(t - k) \qquad (2.51)$$

These iterations can be done as many times as required to obtain $\varphi(t)$ over a grid of dyadic points. An unique set of φ can be obtained through normalization using Equation 2.39. MATLAB program 2.2 shows the derivation of Daubechies scaling functions and wavelets using wavelet toolbox functions.

The scaling functions and the wavelets functions for Haar wavelets (D2) and Daubechies D4 and D12 are presented in Figure 2.2

2.3.3 Discrete wavelet transform

The *Discrete Wavelet Transform* (DWT) is the multi-resolution representation of a finite length discretized signal from fine to coarse scale. For time signals, each of these scales represent a certain frequency band with a time resolution matching the frequency content. Higher frequency components have

```
---------------------------------------------------
MATLAB program 2.1

---------------------------------------------------
% Method 1
dbM='db2'; % for D4
% M=N/2, where, N=2, 4, 6 so on
a_k=dbwavf(dbM); % a_k are the filter coefficients
% a_k are normalized as sum(a_k)=1
%
% Method 2
sum_a_k=1; % sum_a_k=sum(a_k)
a_k=dbaux(M,sum_a_k);
%
% a_k can also be normalized as sum(a_k)=2,
% according to Equation 2.40
% Values in Table 2.2 are normalized as sum(a_k)=2
```

```
---------------------------------------------------
MATLAB program 2.2

---------------------------------------------------
dbM='db2'; % for D4
% M=N/2, where, N=2, 4, 6 so on

iter=3; [phi,psi,xval]=wavefun(dbM,iter);
% phi : scaling function, psi : wavelet function
%
% Normalization according to Equation 2.39

% total number of iteration = 2^iter
% Iteration 1: evaluates phi & psi at integer
% points in [0 2M-1]
% Iteration 2: evaluates phi & psi at mid-points
%  of the integer points
% Iteration 3: evaluates phi & psi at mid-points
% of the grid points formed by Iterations 1 & 2

% xval : grid points created in [0 2M-1]

% Figure 2.4 plot the scaling and wavelet functions
% for,D2, D4 and D12 using this program.

% Plot of scaling function
Figure(1) plot(XVAL,PHI)

% Plot of wavelet function
Figure(2) plot(XVAL,PSI)
```

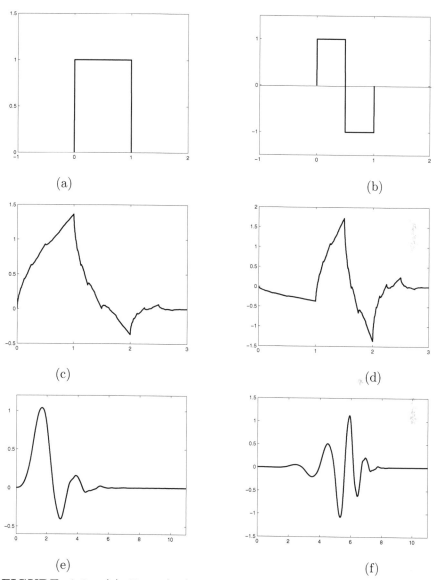

FIGURE 2.2: (a) Haar (D2) scaling function, (b) Haar wavelet, (c) Daubechies D4 scaling function, (d) Daubechies D4 wavelet, (e) Daubechies D12 scaling function, and (f) Daubechies D12 wavelet

a finer time resolution while lower frequency components have a coarser time resolution. DWT can be performed using a fast algorithm referred to as Mallat's transform [57]. Mallat's transform follows a natural tree decomposition and transfers data from one level of resolution j, to the coarser level of resolution $j - 1$, while inverse Mallat's transform follows the reverse process. As given by Equation 2.31, $c_{j,k}$ are the approximation coefficients or the scaling function coefficients of the function $F(t)$ at the resolution level j. Similarly, $d_{j,k}$ are the details or wavelet coefficients of $F(t)$ at the resolution level j as given by Equation 2.36. Equations 2.31 and 2.36 can be applied for discretized form of $F(t)$ given as

$$F_i = F(i\Delta t) \quad \text{where } \Delta t \text{ is the time sampling rate}$$
$$i = 0, 1, \ldots, n - 1$$

In summary,

1. The forward Mallat's transform or Mallat's decomposition decomposes the approximation coefficients $c_{j,k}$ at resolution level j into the approximation coefficients $c_{j-1,k}$ and details coefficients $d_{j-1,k}$ of the next coarse resolution level $j - 1$. This can be repeated for further coarser levels.

2. The inverse Mallat's transform or Mallat's reconstruction reconstructs the approximation coefficients $c_{j,k}$ from $c_{j-1,k}$ and $d_{j-1,k}$ of the resolution level $j - 1$.

The Mallat's forward algorithm is implemented using discrete filters g and h termed as low pass and high pass filters that are convolved with the approximation coefficients $c_{j,k}$ to obtain $c_{j-1,k}$ and $d_{j-1,k}$ after down-sampling by two. The down-sampling is done by keeping only every other sample. Figure 2.3 shows the process of forward Mallat's transform. For Daubechies wavelet, assuming the discretized function F_i to be periodic, the low pass and high pass filters are given as

$$g = [a_0, 0, 0, \ldots, 0, a_{N-1}, \ldots, a_2, a_1]^T \tag{2.52}$$
$$h = [a_{N-1}, 0, 0, \ldots, 0, -a_0, \ldots, a_{N-3}, -a_{N-2}]^T \tag{2.53}$$

where a_k, $k = 0, 1, \ldots, N - 1$ are the filter coefficients as explained in Equation 2.29 and N is the order of Daubechies wavelet as in Equation 2.43.

Figure 2.4 shows the inverse Mallat's algorithm or Mallat's reconstruction. Here, the approximation coefficients $c_{j,k}$ at the higher or finer resolution level j are reconstructed from the approximation $c_{j-1,k}$ and detail $d_{j-1,k}$ coefficients of the coarser level $j-1$. Here, $c_{j-1,k}$ and $d_{j-1,k}$ are convolved with low pass g_1 and high pass h_1 filters after up-sampling by two. The up-sampling is done by adding zero in between two samples of $c_{j-1,k}$ and $d_{j-1,k}$. $c_{j,k}$ are reconstructed by adding the convolved data. Similar to Mallat's decomposition, the low pass

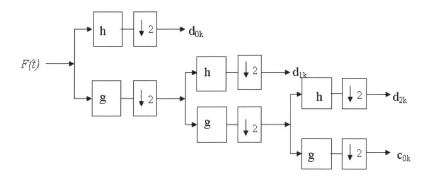

FIGURE 2.3:   Three level forward Mallat's transform

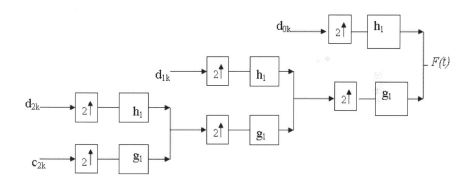

FIGURE 2.4:   Three level inverse Mallat's transform

and high pass discrete filters for Mallat's reconstruction using Daubechies wavelets and assuming F_i to be periodic are as follows,

$$g_1 = [a_{N-1}, -a_{N-2}, a_{N-3}, -a_{N-4}, \ldots, a_1, -a_0, 0, \ldots, 0, 0]^T \quad (2.54)$$

$$h_1 = [a_o, a_1, a_2, a_3, \ldots, a_{N-2}, a_{N-1}, \ldots, 0, 0]^T \quad (2.55)$$

A discretized impulse time signal $F(t)$ is shown in Figure 2.5(a) which has a duration of 50 μs and starts at 100 μs. The time sampling rate for the signal is $\Delta t = 1$ μs and the time window considered is $T_w = 512$ μs. Thus, the total number of sampling (data) points in $F(t)$ is $n = 512$. Figure 2.5(b) shows the details coefficients $d_{j,k}$, at different resolution scales or levels denoted by j. The transformation is done using Haar wavelets and Mallat's decomposition algorithm. The corresponding Matlab functions for Mallat's decomposition and reconstruction using Haar wavelets are given in Matlab Program 2.3 and 2.4.

The maximum number of levels that a signal of length n can be decomposed are $j = 0$ to $n_0 - 1$ and $n = 2^{n_0}$. In addition, there is a level, often referred as level -1. This level contains a single coefficient which represents the average level of $F(t)$, while all other levels have zero mean. It is important to consider this level for proper reconstruction of the signal from the details coefficients. Again, for a given j, k varies from 0 to $2^j - 1$. Next, the decomposed components of $F(t)$ at each level are plotted in Figure 2.6. As mentioned earlier, for a time signal, each of these levels denotes a certain frequency band. These components are obtained by reconstructing the signal from $d_{j,k}$ of each level [74]. The original signal can also be obtained by direct summation of these components. The reconstruction process is shown in Figure 2.7.

A common way of presenting the wavelet coefficients is the wavelet map. It is a very convenient representation of the time-frequency distribution of a time signal. Figure 2.8 shows the wavelet map of impulse load $F(t)$ and, as said earlier, the total number of levels are $j = 0$ to 8, each representing a frequency band covering the range 0 to 500 kHz. The map presented in the figure plots the absolute of wavelet detail coefficient and here a higher order of Daubechies wavelet with $N = 8$ is used for the decomposition. The map has been normalized to a maximum value of unity. The Matlab Program for generation of the wavelet map is given below. The code uses Matlab's in-built function to perform the decomposition.

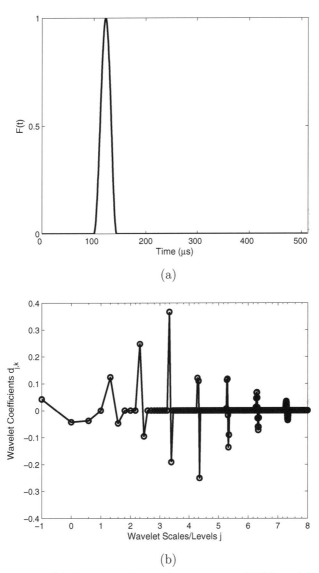

FIGURE 2.5: (a) Discretized impulse time signal F(t) and (b) details coefficients $d_{j,k}$ using Haar wavelets

```
--------------------------------------------------
 MATLAB Program 2.3
--------------------------------------------------
% INPUT SIGNAL
fname=input('Enter input file name: ','s');
pulse=load(fname);
t=pulse(:,1); % Time
F=pulse(:,2); % F(t)
l1=length(F);
%
% Zero padding to make the length: 2^(n0)
n0=9; n=2^n0; F(l1+1:n)=zeros(n-l1,1);
%
% Time Axis
t_s=t(2)-t(1); % Time sampling rate
tf=(n-1)*t_s; % Final time
t=0:t_s:tf;
%
% Forward Haar wavelet transform
d_h=haar_de(F);
%haar_de.m : function for forward Haar wavelet transform
figure(1)
plot(d_h,'r'),% Plot of wavelet coefficients
%
% Inverse Haar wavelet transform
f_r=haar_re(d_h);
%haar_re.m : function for inverse Haar wavelet transform
figure(2)
plot(f_r,'r'),% Plot of reconstructed signal
%
% Reconstruction at each level
d_h_level=zeros(1,n); % Wavelet coefficients at each level
d_h_level(1)=d_h(1); A(1,:)=haar_re(d_h_level);
for i=2:1:n0+1,
d_h_level=zeros(1,n);
d_h_level(2^(i-2)+1:2^(i-1))=d_h(2^(i-2)+1:2^(i-1));
A(i,:)=haar_re(d_h_level); end
%
% Plots of the reconstructed signal at each level
for i=1:1:n0+1, figure(i+2); plot(t,(A(i,:)),'r'), end
%
% Original reconstructed signal obtained by adding the
% signals at each level
figure(n0+4) plot(t,sum(A(1:size(A,1),:)),'r');
```

```
-----------------------------------------------------
 MATLAB Program 2.4
-----------------------------------------------------
function a=haar_de(f)
n=length(f); n0=round(log(n)/log(2));
%
a_k=[1 1]; % Filter coefficients a_k
% for Haar wavelets
b_k=[-1 1]; % Filter coefficients b_k
a=f;
%
for k=n0:-1:1, m=2^(k-1); x=[0]; y=[0];
    for i=1:1:m,
      for j=1:2,
      k(j)=2*i-2+j;
        while k(j)>2*m,
        k(j)=k(j)-2*m;
        end
      end
    z=a(k);
    [mr,nc]=size(z);
        if nc>1,
        z=z.';
        end
    x(i)=a_k*z;
    y(i)=b_k*z;
    end
x=x/2; y=y/2; a(1:m)=x; a(m+1:2*m)=y;
end
```

```
--------------------------------------------------
  MATLAB Program 2.5
--------------------------------------------------
function f=haar_re(a)
n=length(a); n0=round(log(n)/log(2));
%
a_k=[1 1]; % Filter coefficients
% a_k for Haar wavelets
b_k=[-1 1]; % Filter coefficients
% b_k for Haar wavelets
%
f(1)=a(1);
%
for k=1:1:n0, m=2^(k-1);
    for i=1:1:m,
    k(1)=m+i;
        if k(1)<m+1,
            k(1)=k(1)+m;
        end
    z=a(k);
    [mr,nc]=size(z);
        if nc>1,
        z=z.';
        end
    x(2*i-1:2*i)=b_k*z;
    zz=f(k-m);
    [mr,nc]=size(zz);
        if nc>1,
        zz=zz.';
        end
    xx(2*i-1:2*i)=a_k*zz;
    end
f(1:2*m)=x+xx;
end
```

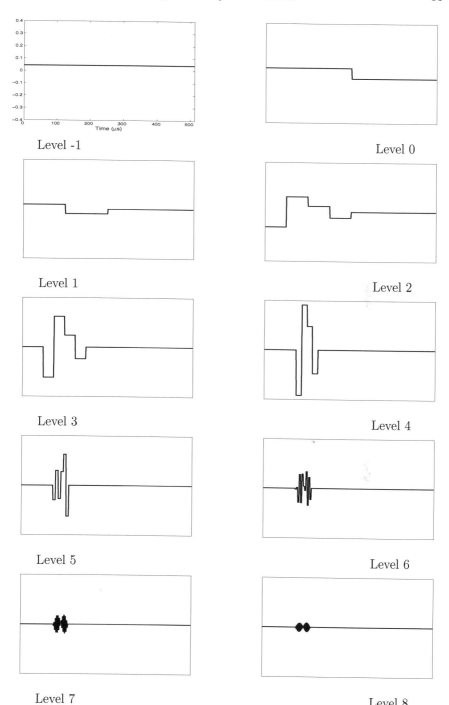

FIGURE 2.6: Components of $F(t)$ shown in Figure 2.5 at each wavelet level $j = 0$ to 8

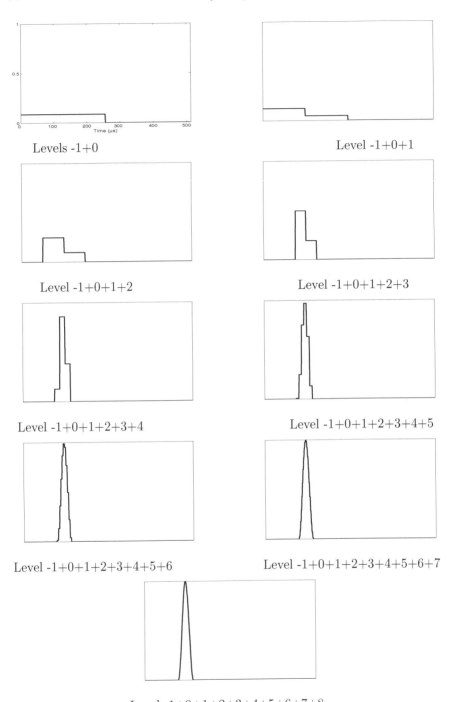

FIGURE 2.7: Reconstruction of $F(t)$ shown in Figure 2.5

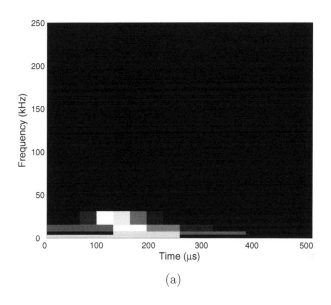

(a)

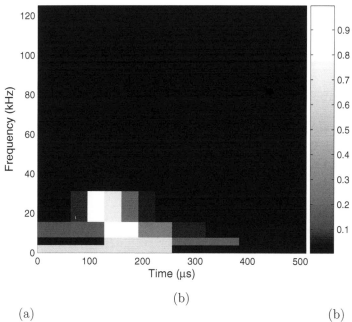

(b)

(a)   (b)

FIGURE 2.8:   (a) Wavelet map for the impulse load $F(t)$ shown in Figure 2.5; (b) wavelet map plotted up to 125 kHz

```
----------------------------------------------------
 MATLAB Program 2.6
----------------------------------------------------
% INPUT SIGNAL
fname=input('Enter input file name: ','s');
pulse=load(fname);
t=pulse(:,1); % Time
F=pulse(:,2); % F(t)
l1=length(F);
% Zero padding to make the length: 2^(n0)
n0=9; n=2^n0; F(l1+1:n)=zeros(n-l1,1);
% Time Axis
t_s=t(2)-t(1); % Time sampling rate
tf=(n-1)*t_s; % Final time
t=0:t_s:tf;
% Frequency Axis
f_max=1/(4*t_s); % Maximum frequency
f_axis=zeros(n0-1,1); f_axis(1,1)=f_max;
for i=2:n0-2
    f_axis(i,1)=f_axis(i-1,1)/2;
end f_axis(n0-1,1)=0;
f_axis=(1e-03)*f_axis; % Frequency axis in kHz
%
% Daubechies Wavelet Decomposition
N=16; % Order of Daubechies Wavelet
L=int2str(N/2);
DW=['db' L];% String for Daubechies wavelet with order N
dwtmode('per'); % Periodic Extension
[C BL]=wavedec(F,n0-1,DW);
% Matlab function for wavelet decomposition
% n0-1 Maximum number of levels of wavelet decomposition
% Matrix for Wavelet Map
WM=zeros(n,n0-1);
for i=1:n0-1
    DC=C(BL(i+1)+1:BL(i+2));
%
    for j=1:2^(n0-i)
    WM(j:2^(n0-i):n,i)=abs(DC(:,1).^2);
    end
end
% Plotting of wavelet map
 figure(1)
 pcolor(t,flipud(f_axis),WM.')
 shading('flat') colormap(pink(128))
 colorbar caxis([0 1.0]);
```

Chapter 3

Structural Dynamics: Introduction and Wavelet Transform

In the last chapter, the concept of wavelet transform in relation with other integral transforms particularly for solution of mechanical dynamics problem has been elaborated. In this chapter we focus on using wavelet transform to solve vibration problems of discrete and continuous structural systems. The example problems considered to explain the wavelet transform-based techniques are the conventional spring-mass-damper system. Prior to the sections on wavelet transform, a brief introduction to the preliminary topics of structural dynamics is provided for completeness of the chapter and also for better understanding to the readers.

As mentioned earlier, the wavelets used throughout this chapter and the book are Daubechies compactly supported wavelets. Though several other forms of wavelets have been used extensively for solution of structural dynamic problem and have been reported in literature, in most of these cases, wavelets have been used as post processing tools either to extract system information from measured dynamic response or to improve the quality of experimentally recorded dynamic response to ease the extraction of different system information. Examples of such applications are de-noising of response signal, extraction of natural frequency and damping parameters from such responses, damage detection, and several others. Though these are important applications of wavelet transform in structural dynamics, this book has slightly more varied objectives. An important focus of the book other than giving a mechanical engineering perspective to wavelet transform and its use for the above mentioned applications is the wavelet-based solution of equations of motion or governing equations associated with vibration and wave propagation problems. For this purpose wavelet functions are used as approximation bases and such functions with localized supports are necessary for solving finite domain problems. In addition to localized support, orthogonality of these functions is also important. Daubechies wavelets satisfy all these criteria among only few other wavelets like bi-orthogonal B-Spline wavelets. Extension of the methods described in this book to the use of other such wavelets may not be direct but sufficiently straightforward.

In this chapter, after the brief introduction to vibrations of discrete structural systems and the conventional analytical solution techniques, the use of wavelet transform for numerical simulation of free and forced vibrations of

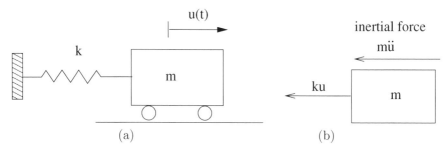

FIGURE 3.1:   (a) SDOF system: undamped spring-mass; (b) free body diagram

multi-degree of freedom (MDOF) system is described in detail. The wavelet-based solution procedure of single-degree of freedom (SDOF) system can be directly derived from that for MDOF systems. Similar method for solution of continuous system has also been explained. However, the methods can be applied for continuous system after they have been discretized through Finite Element (FE) or other methods.

3.1   Free vibration of single degree of freedom systems

In this Section, free vibration of a single degree of freedom (SDOF) system is revisited and analyzed without going into much detail. Though vibration of an SDOF is preliminary, several concepts of structural dynamics and vibration are best understood with the example of an SDOF system particularly spring-mass-damper system. The analysis techniques for vibration of MDOF system are sometimes a straightforward extension of those for SDOF systems. In addition, the wavelet method implemented for simulation of time response can be explained in a most simplified and efficient way using the example of an SDOF system. Again, implementation of the wavelet method for solution of MDOF systems is a direct extension of that for a SDOF system.

The section topic can be started by considering an undamped spring-mass system shown in Figure 3.1(a). The first step in obtaining the response of such structure is to derive the equation of motion (EOM). EOM is the mathematical expression defining the dynamic displacements of a structural system, solution of which gives the complete description of the structure as a function of time [60]. One of the efficient and generalized methods to derive EOM of a structural system is through Lagrange equation using Hamilton's principle. In this Section, however, we will confine ourselves to deriving the EOM from dynamic equilibrium using D'Alembert's principle. The principle states that a mass under motion experiences an inertial force proportional to its acceleration and opposing the motion. Thus the free body diagram of the mass in the spring-mass system shown in Figure 3.1(a) under dynamic equilibrium can be written as

$$m\ddot{u} + ku = 0 \tag{3.1}$$

The free body diagram is also shown in Figure 3.1(b). Equation 3.1 can rewritten in the form

$$\ddot{u} + \omega_n^2 u = 0 \quad \text{where,} \quad \omega_n^2 = \frac{k}{m} \tag{3.2}$$

Here ω_n is the natural frequency of the SDOF system. Equation 3.2 has a solution of the form

$$\begin{aligned} u(t) &= C_1 e^{i\omega_n t} + C_2 e^{-i\omega_n t} \quad \text{where,} \quad i = \sqrt{-1} \\ &= (C_1 + C_2)\cos(\omega_n t) + i(C_1 - C_2)\sin(\omega_n t) \end{aligned} \tag{3.3}$$

Applying initial conditions, $u(0) = u_0$ and $\dot{u}(0) = \dot{u}_0$ in the Equation 3.3, we get the response as

$$u(t) = u_0 \cos(\omega_n t) + \frac{\dot{u}_0}{\omega_n} \sin(\omega_n t) \tag{3.4}$$

Further, the response given by Equation 3.4 can be written in the following form by substituting $u_0 = A\cos\phi$ and $\dot{u}_0 = A\sin\phi$,

$$u(t) = A\cos(\omega_n t - \phi) \tag{3.5}$$

$$A = \sqrt{u_0^2 + \left(\frac{\dot{u}_0}{\omega_n}\right)^2} \quad \text{and} \quad \tan\phi = \frac{\dot{u}_0}{u_0\omega_n}$$

Thus, it can be concluded that the free vibration response of undamped SDOF is oscillatory with frequency ω_n. The amplitude and the phase angle, however, depend on the initial excitation.

Next, damping is added to the spring-mass SDOF system analyzed so far. Linear viscous damping is considered here to explain the preliminary effects of damping on free vibration of a SDOF system. The spring-mass-damper system and the corresponding free body diagram at dynamic equilibrium is shown in Figure 3.2(a) and (b). Similar to the un-damped SDOF system, the EOM of the damped SDOF system can be written from the free body diagram shown in Figure 3.2(b) and is given by

$$m\ddot{u} + c\dot{u} + ku = 0 \tag{3.6}$$

where c is the viscous damping constant. Equation 3.6 can be rewritten as

$$\ddot{u} + 2\xi\omega_n\dot{u} + \omega_n^2 u = 0 \tag{3.7}$$

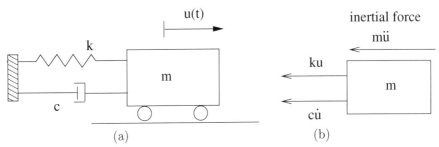

FIGURE 3.2: (a) Damped SDOF system: spring-mass-damper; (b) free body diagram

and $\xi = \frac{c}{2\omega_n m}$ is referred to as the viscous damping factor or damping ratio. Assuming $u(t) = ce^{st}$ as a solution of Equation 3.7 and substituting it into the equation, we get

$$s^2 + 2\xi\omega_n s + \omega_n^2 = 0$$

$$\Rightarrow s = \frac{-2\xi\omega_n \pm \sqrt{4\xi^2\omega_n^2 - 4\omega_n^2}}{2}$$

$$= \left(-\xi \pm \sqrt{\xi^2 - 1}\right)\omega_n \tag{3.8}$$

Therefore, we get the expression for response as

$$u(t) = C_1 e^{(-\xi+\sqrt{\xi^2-1})\omega_n t} + C_2 e^{(-\xi-\sqrt{\xi^2-1})\omega_n t}$$

$$= \left[C_1 e^{\sqrt{(\xi^2-1)}\omega_n t} + C_2 e^{-\sqrt{(\xi^2-1)}\omega_n t}\right] e^{-\xi\omega_n t} \tag{3.9}$$

The nature of the response given by Equation 3.9 is however determined by the value of the damping ratio ξ. Depending on the value of ξ, the SDOF system and the corresponding response can be classified into three cases, namely, *under-damped*, *critically damped*, and *over-damped* systems or responses. For the under-damped case, $0 < \xi < 1$ and s are complex numbers given as

$$s = \left(-\xi \pm i\sqrt{1-\xi^2}\right)\omega_n$$

Hence, the response given by Equation 3.9 will be oscillatory and can be written as

$$u(t) = \left(C_1 e^{i\sqrt{1-\xi^2}\omega_n t} + C_2 e^{-i\sqrt{1-\xi^2}\omega_n t}\right) e^{-\xi\omega_n t}$$

$$= \left(C_1 e^{i\omega_d t} + C_2 e^{-i\omega_d t}\right) e^{-\xi\omega_n t}$$

$$= \left[(C_1 + C_2)\cos(\omega_d t) + i(C_1 - C_2)\sin(\omega_d t)\right] e^{-\xi\omega_n t} \tag{3.10}$$

where $\omega_d = \sqrt{1-\xi^2}\omega_n$ and is referred as *damped* natural frequency. The above analysis shows that the response of an under-damped system will be oscillatory with an exponential decay in the amplitude governed by the term $e^{-\xi\omega_n t}$. Applying the initial conditions, $u(0) = u_0$ and $\dot{u}(0) = \dot{u}_0$, we get the response as

$$u(t) = \left[u_0 \cos(\omega_d t) + \left(\frac{\dot{u}_0 + \xi\omega_n u_0}{\omega_d} \right) \sin(\omega_d t) \right] \tag{3.11}$$

Substituting $u_0 = A\cos\phi$ and $\frac{(\dot{u}_0 + \xi\omega_n u_0)}{\omega_d} = A\sin\phi$, Equation 3.11 can also be expressed as

$$u(t) = A\cos(\omega_d t - \phi), \text{ where } A = \sqrt{u_0^2 + \frac{(\dot{u}_0 + \xi\omega_n u_0)^2}{\omega_d^2}}, \tan\phi = \frac{\dot{u}_0 + \xi\omega_n u_0}{u_0\omega_d} \tag{3.12}$$

An under-damping is the most common form of damping experienced in real life structures. The second case is the critically damped condition with $\xi = 1$ for which s given by Equation 3.8 is real and equal. Thus, the response can be written as

$$u(t) = (C_1 + C_2 t)e^{-\omega_n t} \tag{3.13}$$

The unknown constants C_1 and C_2 can be obtained from the initial condition as done for the under-damped case. Equation 3.13 shows that the response at critical damping is non-oscillatory and the amplitude will decrease exponentially. Critical damping can also be thought as the minimum damping required to prevent oscillatory motion. It can be seen from the following derivation that even for over-damped system, the response is non-oscillatory, but the rate of decay of the response is lower than the critically damped system. For over-damped system with $\xi > 1$, the response is obtained as

$$u(t) = (C_1 e^{\omega_d t} + C_2 e^{-\omega_d t})e^{-\xi\omega_n t} \tag{3.14}$$

Figure 3.3 plots and compares the free vibration responses of SDOF under different damping conditions. The SDOF has a natural frequency of 4.0 Hz and the initial conditions are $u_0 = 1$ mm and $\dot{u}_0 = 5$ mm/s.

3.2 Forced vibration of SDOF system

This Section provides a brief overview of the different commonly used analytical techniques to solve the forced vibration problem. Even here, the spring-mass-damper system explained in the previous Section is used as an example of SDOF system. Before going into the details of a generalized solution scheme for a harmonic, periodic, and arbitrary loading condition, a simple example of step loading is dealt with to preview forced vibration of SDOF system.

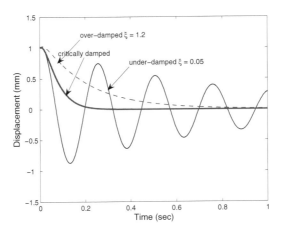

FIGURE 3.3: Free vibration of SDOF system under different damping conditions

FIGURE 3.4: Step load

A step function of magnitude P_0 is shown in Figure 3.4 and the EOM of a spring-mass-damper system under this load is given as

$$m\ddot{u} + c\dot{u} + ku = P_0$$
$$\ddot{u} + 2\xi\omega_n\dot{u} + \omega_n^2 u = \frac{P_0}{m} \tag{3.15}$$

The solution of the above linear, non-homogeneous differential equation is given as

$$u(t) = u_c(t) + u_p(t) \tag{3.16}$$

where u_c is the complementary solution and for under-damped system is of the form given by Equation 3.10. u_p is the particular solution of Equation 3.15 and is given as,

$$u_p(t) = \frac{P_0}{m\omega_n^2} \tag{3.17}$$

Therefore, the response of the SDOF system due to the step loading is given

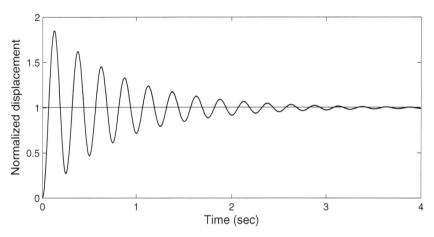

FIGURE 3.5: Response of under-damped SDOF to step loading

as

$$u(t) = \left[(C_1 + C_2)\cos(\omega_d t) + i(C_1 - C_2)\sin(\omega_d t)\right] e^{-\xi\omega_n t} + \frac{P_0}{m\omega_n^2} \qquad (3.18)$$

Under zero initial condition the response is obtained as

$$u(t) = \frac{P}{m\omega_n^2} \left[1 - (\cos\omega_d t + \frac{\xi}{\sqrt{1-\xi^2}}\sin\omega_d t)e^{-\xi\omega_n t}\right] \qquad (3.19)$$

From Equation 3.19, it can be interpreted that the response of under-damped SDOF to step function will be oscillatory about $\frac{P_0}{m\omega_n^2}$ and the response normalized by $\frac{P_0}{m\omega_n^2}$ is plotted in Figure 3.5.

3.3 Harmonic loading

A structure commonly encounters harmonic load from rotating machines, moving loads, and others. Harmonic loading is also used for experimental natural frequencies, damping, and other dynamic parameters. In this Section, the response of damped SDOF to harmonic force is reviewed briefly. Considering a spring-mass-damper system under such loading, the EOM is as follows,

$$m\ddot{u} + c\dot{u} + ku = F_0 \cos\omega t = kA\cos\omega t$$
$$\ddot{u} + 2\xi\omega_n\dot{u} + \omega_n^2 u = \omega_n^2 A\cos\omega t \qquad (3.20)$$

where F_0 is the amplitude of the applied load, which can be arbitrarily represented as $F_0 = kA$. ω is referred as the *driving frequency*. Similar to the problem of step load, the solution of Equation 3.20 is also the superposition of complementary and particular solutions given by Equation 3.16. The complementary solution, u_c, is the same as Equation 3.10. For damped system, u_c will decay towards zero with time and hence it is called a *transient* response. The particular solution, u_p, is however assumed as

$$u_p = X \cos(\omega t - \phi) \qquad (3.21)$$

u_p does not have any damping term and will persist as long as the force acts. This part of the response is called the *steady-state* response. Emphasis is mostly given in understanding the steady-state response since for realistic damping the transient response becomes negligibly small within a short duration. Substituting Equation 3.21 in Equation 3.20 and equating the coefficients of $\cos \omega t$ and $\sin \omega t$, we get

$$X \left[(\omega_n^2 - \omega^2) \cos \phi + 2\xi \omega_n \omega \sin \phi \right] = \omega_n^2 A \qquad (3.22)$$
$$X \left[(\omega_n^2 - \omega^2) \sin \phi - 2\xi \omega_n \omega \cos \phi \right] = 0 \qquad (3.23)$$

Multiplying Equations 3.22 and 3.23 by $\cos \phi$ and $\sin \phi$ respectively followed by addition, we get

$$X(\omega_n^2 - \omega^2) = \omega_n^2 A \cos \phi \qquad (3.24)$$

Similarly, multiplying Equations 3.22 and 3.23 by $\sin \phi$ and $\cos \phi$ respectively followed be substraction, we get

$$X(2\xi \omega_n \omega) = \omega_n^2 A \sin \phi \qquad (3.25)$$

Solving Equations 3.24 and 3.25, the amplitude X and phase angle ϕ are obtained as

$$\frac{X}{A} = H(\omega) = \frac{1}{\left[\left\{ 1 - \left(\frac{\omega}{\omega_n} \right)^2 \right\}^2 + \left(2\xi \frac{\omega}{\omega_n} \right)^2 \right]^{\frac{1}{2}}} \qquad (3.26)$$

$$\tan \phi = \frac{2\xi \frac{\omega}{\omega_n}}{1 - \left(\frac{\omega}{\omega_n} \right)^2} \qquad (3.27)$$

Hence, the steady-state response of a under-damped SDOF is also harmonic with constant amplitude X and phase angle ϕ given by Equations 3.26 and 3.27. The ratio $\frac{X}{A}$ is referred to as *magnification factor* and represented by $H(\omega)$. $H(\omega)$ is however not defined for un-damped vibration at resonance. The magnification factors for different damping ratios are plotted in Figure 3.6. The driving frequency ω corresponding to the peak value of $H(\omega)$ can be obtained by taking the derivative of the expression of $H(\omega)$ given by Equation 3.26 and equating it to zero. This gives $\omega = \omega_n \sqrt{1 - 2\xi^2}$ for which

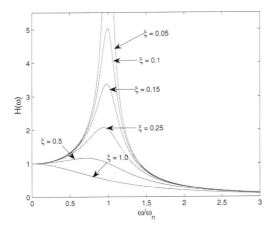

FIGURE 3.6: Magnification factor $H(\omega)$

one gets the peak value of $H(\omega)$ referred to as *quality factor*, Q. Substituting this value of ω in Equation 3.26 and assuming ξ to be small, we get

$$Q \approx \frac{1}{2\xi} \qquad (3.28)$$

Thus, $H(\omega)$ has the maximum value at frequencies slightly lower than the resonance frequency, but this converges to the resonance frequency with decreasing ξ. Such peak value does not exist for $\xi > 1/\sqrt{2}$. These features of $H(\omega)$ can also be observed from Figure 3.6. Next, the variation of phase angle ϕ with ω/ω_n for different values of ξ is plotted in Figure 3.7. The main observation from Figure 3.7 and also the expression of ϕ given by Equation 3.27 is that ϕ is $\pi/2$ at resonance frequency irrespective of ξ.

3.3.1 Periodic excitation

In this Section, the Fourier series-based method to obtain the response of a SDOF to general periodic loading is explained. For the applied load $F(t)$ being periodic, it can be represented by a convergent series of harmonic function as

$$F(t) = \frac{1}{2}a_0 + \sum_{n=1}^{N}(a_n \cos \omega_n t + b_n \sin \omega_n t) \qquad (3.29)$$

$$\text{where,} \quad \omega_n = \frac{2\pi n}{T}, \quad n = 0, 1, 2, \ldots$$

T being the time period of the function. The above equation is similar to Equation 2.19, except that the upper limit of the summation is a finite (N) instead of being infinite. a_n and b_n are the Fourier series coefficients and

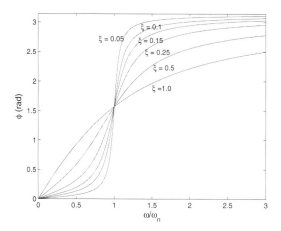

FIGURE 3.7: Phase angle ϕ

are given by Equation 2.20. Several other properties of Fourier series are also explained in Subsection 2.2.2 of Chapter 2. N in Equation 3.29 is chosen either from a convergence study or according to the frequency content of the applied harmonic loading. In the latter method, the maximum value of N should be such that $\omega_n = \frac{2\pi N}{T}$ is greater than the maximum frequency content of $F(t)$. As explained before in Chapter 2, the frequency content of a function or load can be known from its Fourier transform.

After $F(t)$ is expressed in terms of harmonics given by Equation 3.29, the steady state response of the SDOF can be obtained for each harmonic using the methods described in Section 3.3. The final response can be derived by adding these responses for the different harmonic loading using the superposition principle.

3.4 Response to arbitrary loading

In the previous Section, the response of a SDOF to harmonic loading and also generalized periodic loading has been studied. In Section 3.4.2, an analytical technique for obtaining the response to an arbitrary loading will be described. The method is referred to as *Convolution* or *Duhamel's Integral* method. Prior to this, however, one needs to understand the concepts of *impulse* and response to such impulse which is generally termed as *impulsive response*.

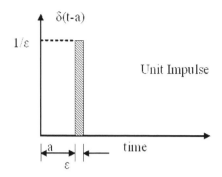

FIGURE 3.8: Unit impulse

3.4.1 Impulse response

In a heuristic sense, an impulsive load is one that acts for a very short duration. Figure 3.8 shows an unit impulse acting at $t = a$. The impulse can be thought to have a width ε, where $\varepsilon \to 0$ and height $1/\varepsilon$. An impulse has the unit of *force-sec*. Mathematically, an unit impulse is represented by *Dirac-delta* function, $\delta(t - a)$, defined as

$$\delta(t - a) = 0 \ \forall \ t \neq a \tag{3.30}$$

$$\int_{-\infty}^{\infty} \delta(t - a)dt = 1 \tag{3.31}$$

The integral in Equation 3.31 being non-dimensional, $\delta(t-a)$ has an unit s^{-1}. An impulse load having arbitrary magnitude of F_0 can be written as

$$F(t) = F_0\delta(t - a) \tag{3.32}$$

where F_0 has the unit of impulse, i.e., *force-sec* like N-s. The EOM of a SDOF under impulse load of magnitude F_0 acting at $t = 0$ can be written as

$$m\ddot{u} + c\dot{u} + ku = F_0\delta(t) \tag{3.33}$$

An impulse load acting at $t = 0$ results in initial velocity and the response due to such a load can be thought of as a free vibration with initial velocity. Applying Newton's law, the initial velocity can be obtained in the following way,

$$F = \frac{mdv}{dt} \tag{3.34}$$

where v is the velocity at any instance of time. Now,

$$F dt = mdv$$
$$\int_0^\varepsilon F dt = \int_0^\varepsilon mdv$$

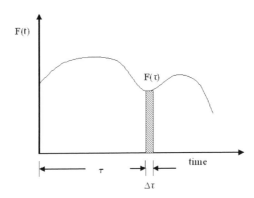

FIGURE 3.9:   Arbitrary loading

$$\int_0^\varepsilon F_0 \delta(t) dt = \int_0^\varepsilon m dv$$

$$F_0 = m[v(\varepsilon) - v(0)] \quad \text{where,} \quad \varepsilon \to 0$$

$$v(\varepsilon) = v(0+) = v(0) = \frac{F_0}{m} \tag{3.35}$$

Thus, substituting the initial conditions, $u(0) = 0$ and $\dot{u}(0) = \frac{F_0}{m}$, into the free vibration response of an under-damped system given by Equation 3.11, we get the response as

$$u(t) = \frac{F_0}{m\omega_d} e^{-\xi\omega_n t} \sin \omega_d t \tag{3.36}$$

Thus the response due to unit impulse is given and defined as

$$h(t) = \frac{1}{m\omega_d} e^{-\xi\omega_n t} \sin \omega_d t \tag{3.37}$$

Similarly, the response to unit impulse load acting at $t = \tau$ can be written as

$$h(t - \tau) = \frac{1}{m\omega_d} e^{-\xi\omega_n(t-\tau)} \sin \omega_d(t - \tau) \quad t > \tau \tag{3.38}$$

3.4.2   Convolution integral method

Consider an arbitrary load as shown in Figure 3.9. The load can be thought of as the superposition of several impulsive loads acting at different time instances. Such an impulsive load acting at $t = \tau$ can be written as

$$F(t) = F(\tau)\Delta\tau\delta(t - \tau) \tag{3.39}$$

where $\Delta\tau$ is the width of the impulse. The response of the under-damped SDOF to this loading can be obtained as

$$u_\tau(t) = F(\tau)\Delta\tau h(t - \tau) \tag{3.40}$$

Equation 3.40 is obtained by replacing F_0 by $F(\tau)\Delta\tau$ in Equation 3.36. Hence, the response due to the entire load shown in Figure 3.9 can be obtained using principle of superposition as

$$u(t) = \sum F(\tau)\Delta\tau h(t - \tau) \qquad (3.41)$$

Again with $\Delta\tau \to 0$, the summation in Equation 3.41 can be replaced by an integral and can be written as

$$u(t) = \int_0^t F(\tau)h(t - \tau)d\tau \qquad (3.42)$$

The integral in Equation 3.42 is termed as a convolution or Duhamel integral.

3.4.3   Direct time integration methods

In the last two Subsections and in Section 3.3 different analytical methods to obtain the forced response of damped SDOF have been explained briefly, mostly for completeness. These analytical methods give a better insight to the physics of the structural behavior. The main constraint of these methods is however that they are applicable only when the loading function is known analytically and hence cannot be applied to real life problems where mostly discrete values of load are known.

The EOM of a damped SDOF being second order linear ODE in time, it can be easily solved by adopting one of the several numerical ODE solvers. For these numerical solution schemes, the loading need not be analytical, rather a loading known analytically is converted to a discrete loading function. These numerical schemes can be subdivided as *direct* and *indirect* time integration schemes. In *direct* time integration, no transformation of the differential equation to a different domain or form is performed.

For direct time integration, we need to know the initial conditions u_0 and $\dot{u}_0$ to proceed. Let 0 to t_f be the time window to be considered and it is discretized at n time sampling points. Let $\tau = 0, 1, 2, \ldots, n - 1$ be the sampling points, then

$$t = \Delta t \tau \qquad (3.43)$$

where Δt is the time interval between two sampling points. The underlying idea of direct time integration scheme is to obtain the response at the sampling points. An approximation function is used to determine the solution at a time instance knowing the solutions at preceding time instances. Some examples of commonly used direct time integration schemes are *finite difference* methods, *Wilson-θ* method, *Newmark's* time integrations, and others. Among these, finite difference methods fall under the class of *explicit* time integration schemes, while Wilson-θ and Newmark's methods are *implicit* time integration schemes. In explicit schemes, the solution at a time instance is obtained using the differential equation at the preceding time instance. In contrast, in implicit schemes, the equation at the same time instance is used. The details of these methods, and their stability and suitability for a particular problem have been discussed in details by Bathe [3] and also in other references.

3.5   Response of SDOF through wavelet transform

In this Section, the implementation of wavelet transform to obtain the forced vibration response of SDOF is explained in detail. The method presented is a numerical scheme similar to the direct time integration techniques explained in the previous Subsection. The wavelet transform-based method is however an *indirect* time integration scheme as the differential equations are solved after transformation to a different domain, wavelet domain in this case. Other such indirect time integrations methods based on Fourier and Laplace transforms are also used for solution of structural dynamics problems. Wavelet transform, however, has certain advantages over the other two transforms mentioned. These have been discussed in Chapter 2.

Here, the wavelet transform-based method is described through the example of a damped SDOF. The method can however be directly extended for solution of MDOF problems and it has been presented in the later Sections on MDOF. Let us consider the SDOF spring-mass-damper system for which the EOM is given as

$$m\ddot{u} + c\dot{u} + ku = F(t) \tag{3.44}$$

Here, only discrete values of $F(t)$ at the time sampling point are required to be known. Even for cases, where the loading function $F(t)$ is known analytically, it needs to discretized at the time sampling points. This is because, as mentioned earlier, the wavelet transform-based method presented here is a numerical scheme and the solution also gives discrete values of the response $u(t)$ at those time sampling points. Let $u(t)$ and $F(t)$ be discretized at n points in the time window $[0 \ t_f]$. Let $\tau = 0, 1, 2, \ldots, n-1$ be the sampling points with Δt being the time interval. The relation between t and τ is given by Equation 3.43. The function $u(t)$ can be approximated by Daubechies scaling function $\varphi(\tau)$ at an arbitrary scale as

$$u(t) = u(\tau) = \sum_{k} u_k \varphi(\tau - k), \quad k \in \mathbb{Z} \tag{3.45}$$

where u_k are the approximation coefficients. Similarly, $F(t)$ can be approximated as

$$F(t) = F(\tau) = \sum_{k} F_k \varphi(\tau - k), \quad k \in \mathbb{Z} \tag{3.46}$$

and F_k are the approximation coefficients of $F(\tau)$. Substituting Equations 3.45, 3.46, and 3.43 into Equation 3.44, we get

$$\frac{1}{\Delta t^2} m \sum_k u_k \varphi''(\tau - k) + \frac{1}{\Delta t} c \sum_k u_k \varphi'(\tau - k) + k \sum_k u_k \varphi(\tau - k) = \sum_k F_k \varphi(\tau - k)$$

(3.47)

Multiplying both sides of Equation 3.48 by $\varphi(\tau - j)$, where, $j = 0, 1, 2, \ldots, n - 1$, and taking inner products, we get

$$\frac{1}{\Delta t^2} m \sum_k u_k \int \varphi''(\tau - k) \varphi(\tau - j) d\tau + \frac{1}{\Delta t} c \sum_k u_k \int \varphi'(\tau - k) \varphi(\tau - j) d\tau$$

$$+ k \sum_k u_k \int \varphi(\tau - k) \varphi(\tau - j) d\tau = \sum_k F_k \int \varphi(\tau - k) \varphi(\tau - j) d\tau \qquad (3.48)$$

Using the orthogonal properties of the translates of scaling function given by Equation 2.41, Equation 3.48 can be written as

$$\frac{1}{\Delta t^2} m \sum_{k=j-N+2}^{j+N-2} \Omega_{j-k}^2 u_k + \frac{1}{\Delta t} c \sum_{k=j-N+2}^{j+N-2} \Omega_{j-k}^1 u_k + k u_j = F_j \quad j = 0, 1, 2 \ldots, n-1$$

(3.49)

where, N is the order of Daubechies wavelet as explained in Section 2.3.2 in Chapter 2. Ω_{j-k}^1 and Ω_{j-k}^2 are the connection coefficients defined as

$$\Omega_{j-k}^1 = \int \varphi'(\tau - k) \varphi(\tau - j) d\tau \qquad (3.50)$$

$$\Omega_{j-k}^2 = \int \varphi''(\tau - k) \varphi(\tau - j) d\tau \qquad (3.51)$$

For Daubechies compactly supported wavelets, Ω_{j-k}^1 and Ω_{j-k}^2 are non-zero only in the interval $k = j - N + 2$ to $k = j + N - 2$. The details for evaluation of connection coefficients for different orders of derivative are given by Beylkin [8]. Here, a MATLAB function (MATLAB Program 3.1) which can be used to derive the connection coefficients for Daubechies scaling functions is present. The function uses an iterative algorithm to derive these parameters.

While dealing with finite length data sequence, problems arise at the boundaries. It can be observed from the Equation 3.49 that certain coefficients u_j near the vicinity of the boundaries ($j = 0$ and $j = n - 1$) lie outside the time window given by $[0 \, t_f]$ and defined by $j = 0, 1, 2 \ldots, n-1$. Several approaches like capacitance matrix methods [79, 80], and penalty function methods for treating boundaries are reported in the literature. For the present problem, first a circular convolution method is adopted assuming periodicity of solution and can be applicable for periodic loading. The solution obtained using this boundary treatment is exactly similar to those obtained using DFT to solve the problem and, thus, cannot simulate the time domain response. It however can be used to obtain the frequency domain response and those details

```
--------------------------------------------------
MATLAB program 3.1
--------------------------------------------------
function [K_omega]=ccoeff(L,n)
% K_omega: Vector containing CC
% L: Order of Daubechies wavelet; n: order of CC (derivative)
% Autocorrelation Coefficients a_2k-1;
M=L/2; a=zeros(1,L-1);
C_M=(factorial(2*M-1)/(factorial(M-1)*(4^(M-1))))^2;
for i=1:2:L-1
m=(i+1)/2;
a(i)=((-1)^(m-1))*C_M/(factorial(M-m)*factorial(M+m-1)*i);
end
% CONNECTION COEFFICIENTS
A1=zeros(L-1,L-1); for l=0:L-2
    if 2*l<=L-2
        i=2*l; A1(l+1,i+1)=2^n;
    end
for k=1:L/2
        i1=2*l-2*k+1;
    if i1<=L-2  & i1>=0
        i=i1;
        A1(l+1,i+1)=A1(l+1,i+1)+(2^(n-1))*a(2*k-1);
    elseif i1>=-L+2 & i1<0
        i=-i1;
        A1(l+1,i+1)=A1(l+1,i+1)+((-1)^n)*(2^(n-1))*a(2*k-1);
    end
    i2=(2*l+2*k-1);
    if i2<=L-2  & i2>=0
        i=i2;
        A1(l+1,i+1)=A1(l+1,i+1)+(2^(n-1))*a(2*k-1);
    elseif i2>=-L+2 & i2<0
        i=-i2;
        A1(l+1,i+1)=A1(l+1,i+1)+((-1)^n)*(2^(n-1))*a(2*k-1);
    end
 end
 end
for g=1:L-1
    A1(g,g)=A1(g,g)-1;
end
ZT=null(A1);
NM=0; % for normalization
for i=2:L-1
    NM=NM+((i-1)^n)*ZT(i);
end
K_omega=((-1)^n)*ZT*factorial(n)/(2*NM);
return
```

have been explained in Chapter 6. The second method is a wavelet-based extrapolation scheme proposed by Amaratunga and Williams [1, 2, 101] and is implemented for solution of the boundary value problem. The method is particularly suitable for approximation in time for the ease of imposition of initial conditions. The details of the formulation is given in the following Subsection.

After treating the boundary for analysis of the finite system, Equation 3.49 can be written as a matrix algebraic equation and can be solved following conventional techniques.

In the following Subsection, as mentioned earlier, the treatment of boundaries for finite domain analysis is presented. Equation 3.49 gives n coupled algebraic equations, which are to be solved for u_j using a method described later. For numerical implementation, we can deal with only finite sequence. In other words, $u(t)$ and hence u_j are only known in the interval $[0 \ t_f]$ and $j = 0$ to $j = n - 1$. In Equation 3.49, some equations corresponding to $j = 0$ to $j = N - 2$ contain coefficients u_j that lie outside $[0 \ t_f]$. Similarly on the other boundary, for $j = (n - 1) - N + 2$ to $j = n - 1$, same problem exists. In this subsection, the details of the approach adopted to solve this boundary value problem are provided.

3.5.1 Non-periodic boundary condition

As said before, here the boundaries are treated using wavelet extrapolation method of Daubechies compactly supported wavelets as proposed by Amaratunga and Williams [101, 1, 2]. In this method a polynomial of order $p - 1$, p generally taken as $N/2$, is assumed to extrapolate the values at the boundaries. Since in the present method, the wavelets are used in time, the unknown coefficients on the LHS (i.e., $u_{-1}, u_{-2}, \ldots, u_{-N+2}$) are extrapolated from the initial values. The coefficients on the RHS are extrapolated from the known coefficients $u_{(n-1)-p+1}, u_{(n-1)-p+2}, \ldots, u_{n-1}$.

Assuming polynomial representation of order $p - 1$ for u in the vicinity of $t = 0$ and using Equation 3.45, we get

$$u(\tau) = \sum_{k} u_k \varphi(\tau - k) = \sum_{l=0}^{p-1} c_l \tau^l \tag{3.52}$$

where c_l are constant coefficients. Taking inner product on both sides of Equation 3.52, we get

$$u_j = \sum_{l=0}^{p-1} c_l \mu_j^l \quad j = -1, -2, \ldots, -N + 2 \tag{3.53}$$

where μ_j^l are the moments of the scaling function defined as

$$\mu_j^l = \int_{-\infty}^{\infty} \tau^l \varphi(\tau - j) d\tau \tag{3.54}$$

and are derived by solving a recursive equation [46]. Solution of Equation 3.52 to obtain c_l requires $p - 1$ initial values of $u(\tau)$ at $\tau = 0, 1, \ldots, p - 1$ and these may be obtained using schemes like finite difference. Next, the values of c_l obtained in terms of the initial values are substituted back into Equation 3.53. Thus, the unknown coefficients u_j, $j = -1, -2, \ldots, -N + 2$ are obtained as

$$
\begin{bmatrix} u_{-1} \\ u_{-2} \\ \vdots \\ u_{-N+2} \end{bmatrix} = \begin{bmatrix} \mu_{-1}^0 & \mu_{-1}^1 & \cdots & \mu_{-1}^{p-1} \\ \mu_{-2}^0 & \mu_{-2}^1 & \cdots & \mu_{-2}^{p-1} \\ \vdots & \vdots & \cdots & \vdots \\ \mu_{-N+2}^0 & \mu_{-N+2}^1 & \cdots & \mu_{-N+2}^{p-1} \end{bmatrix} \begin{bmatrix} c_0 \\ c_1 \\ \vdots \\ c_{p-1} \end{bmatrix} \tag{3.55}
$$

The unknown coefficients at the RHS boundary are evaluated assuming the same polynomial representation and

$$
u_j = \sum_{l=0}^{p-1} c_l \mu_{j-n}^l \qquad j = (n-1) - p + 1, \ (n-1) - p + 2, \ldots, n - 1 \tag{3.56}
$$

Equation 3.56 can be written in matrix form as

$$
\begin{bmatrix} \mu_{-p}^0 & \mu_{-p}^1 & \cdots & \mu_{-p}^{p-1} \\ \mu_{-p+1}^0 & \mu_{-p+1}^1 & \cdots & \mu_{-p+1}^{p-1} \\ \vdots & \vdots & \cdots & \vdots \\ \mu_{-1}^0 & \mu_{-1}^1 & \cdots & \mu_{-1}^{p-1} \end{bmatrix} \begin{bmatrix} c_0 \\ c_1 \\ \vdots \\ c_{p-1} \end{bmatrix} = \begin{bmatrix} u_{(n-1)-p+1} \\ u_{(n-1)-p+2} \\ \vdots \\ u_{(n-1)} \end{bmatrix} \tag{3.57}
$$

The c_l obtained are then substituted into Equation 3.56 for $j = n, \ n + 1, \ldots, n + N - 2$ to derive $u_{(n-1)-p+1}, \ u_{(n-1)-p+2}, \ldots, u_{n-1}$ as

$$
\begin{bmatrix} u_n \\ u_{n+1} \\ \vdots \\ u_{n-1+N-2} \end{bmatrix} = \begin{bmatrix} \mu_0^0 & \mu_0^1 & \cdots & \mu_0^{p-1} \\ \mu_1^0 & \mu_1^1 & \cdots & \mu_1^{p-1} \\ \vdots & \vdots & \cdots & \vdots \\ \mu_{-N+2}^0 & \mu_{-N+2}^1 & \cdots & \mu_{-N+2}^{p-1} \end{bmatrix} \begin{bmatrix} c_0 \\ c_1 \\ \vdots \\ c_{p-1} \end{bmatrix} \tag{3.58}
$$

Finally, these coefficients are substituted in Equation 3.49 and the system of coupled equations can be written in a matrix form as

$$
(m\Gamma^2 + c\Gamma^1 + kI)\{u_j\} = \{F_j\} \tag{3.59}
$$

It should be noted that the connection coefficient matrices Γ^1 and Γ^2 are independent of the problem and depend only on order of the Daubechies wavelet used, i.e., N.

It can be seen from the above derivation that the wavelet coefficients of the first and second order derivative can be obtained as

$$
\{\dot{u}_j\} = \Gamma^1 \{u_j\} \tag{3.60}
$$

$$\{\ddot{u}_j\} = \Gamma^2\{u_j\} \tag{3.61}$$

The second derivative can also be written as

$$\{\ddot{u}_j\} = \Gamma^1\{\dot{u}_j\} \tag{3.62}$$

Substituting Equation 3.60 in Equation 3.62 we get

$$\{\ddot{u}_j\} = [\Gamma^1]^2\{u_j\} \tag{3.63}$$

Thus though the second order connection coefficient matrices Γ^2 can be evaluated independently [8], they can also be written as

$$\Gamma^2 = [\Gamma^1]^2 \tag{3.64}$$

Substituting Equation 3.64 into Equation 3.59, we get

$$(m\Gamma^{1^2} + c\Gamma^1 + kI)\{u_j\} = \{F_j\} \tag{3.65}$$

Equation 3.65 can be solved to obtain $\{u_j\}$ using one of the conventional matrix solution techniques like Gauss elimination, LU factorization, and others.

Figure 3.10 presents and compares the response of a spring-mass-damper system to periodic square simulated using the wavelet method and Runge-Kutta 45 ODE solver. For the latter method, Matlab function *ode45* has been used. The SDOF system considered has a natural frequency of 4.0 Hz and damping ratio 0.02 with unit mass. The applied load is shown in the inset of Figure 3.10 and has a period of 0.01 s. The time sampling rate considered for both the simulations is 0.5 ms. As it can be seen from the plots, the responses are exactly similar. It should be mentioned here that such time domain responses cannot be obtained using other commonly used integral transforms like Fourier and Laplace transforms. In addition, wavelet method applied for the SDOF problem is computationally as efficient as the other direct time integration scheme. One advantage of the wavelet method is that it can be used for inverse problems of force reconstruction from the measured response. The wavelet method also directly gives the frequency domain information which are essential in several dynamic problems. This advantage of the method is however not prominent for the present case of SDOF considered and will be explained for the examples of MDOF systems and wave propagation in continuous waveguides later. Next, responses of the above system to a unit pulse loading are presented in Figure 3.11. All the other parameters are kept the same except the time sampling rate which is 2.0 ms now. The pulse has a width of 50 ms and is shown as inset in Figure 3.11. This form of loading is used to exhibit the efficiency of the wavelet method in handling non-periodic, transient loading conditions.

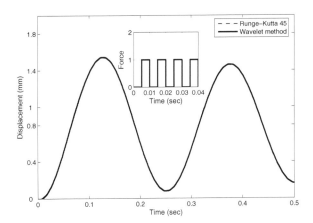

FIGURE 3.10: Response of a spring-mass-damper system to periodic square loading (inset)

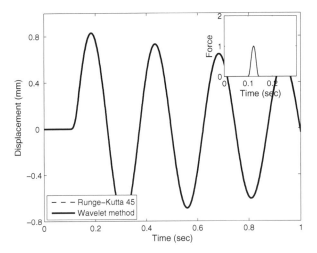

FIGURE 3.11: Response of a spring-mass-damper system to transient pulse loading (inset)

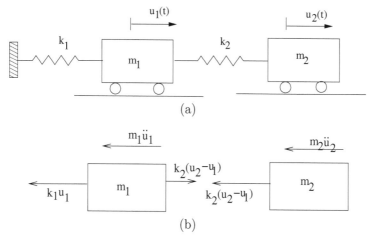

(a)

(b)

FIGURE 3.12: (a) A two-DOF spring-mass system; (b) free body diagrams

3.6 Free vibration of multi-degree of freedom system

The essential difference between the free vibration responses of an SDOF and an MDOF is that for the latter case, the vibration is a linear combination of several *normal modes* of vibration. The number of such normal modes is equal to the number of degrees of freedom in the MDOF system. Each of this normal mode of vibration is governed by a natural frequency and a *mode shape*. The concept of mode shape comes from the property of normal mode, where the deformation shape of the system undergoing the vibration remains unchanged with time. This can be explained in other words as the ratio of displacement of the degrees of freedom remains constant for normal mode of vibration. This Section provides a brief overview of free vibration of an MDOF system.

Considering a spring-mass system with two-DOF as shown in Figure 3.12, the equation of motion can be written as

$$m_1\ddot{u}_1 + k_1 u_1 - k_2(u_2 - u_1) = 0 \qquad (3.66)$$

$$m_2\ddot{u}_2 + k_2(u_2 - u_1) = 0 \qquad (3.67)$$

Equations 3.66 and 3.67 can be written in matrix form as

$$\begin{bmatrix} m_1 & 0 \\ 0 & m_2 \end{bmatrix} \begin{Bmatrix} \ddot{u}_1 \\ \ddot{u}_2 \end{Bmatrix} + \begin{bmatrix} (k_1 + k_2) & -k_2 \\ -k_2 & k_2 \end{bmatrix} \begin{Bmatrix} u_1 \\ u_2 \end{Bmatrix} = 0 \qquad (3.68)$$

Assuming the solution of Equation 3.68 for $u_1(t)$ and $u_2(t)$ as $A_1 \cos(\omega t - \varphi)$ and $A_2 \cos(\omega t - \varphi)$, and substituting it back into Equation 3.68, we can solve for ω, A_1, A_2. It should be noted that the above assumption satisfies the condition for normal mode of vibration as the ratio of $u_1(t)$ and $u_2(t)$ will remain unchanged with time. Next, substituting the assumed solutions into Equation 3.68, we get

$$-\omega^2 \begin{bmatrix} m_1 & 0 \\ 0 & m_2 \end{bmatrix} \begin{Bmatrix} A_1 \\ A_2 \end{Bmatrix} + \begin{bmatrix} (k_1 + k_2) & -k_2 \\ -k_2 & k_2 \end{bmatrix} \begin{Bmatrix} A_1 \\ A_2 \end{Bmatrix} = 0 \qquad (3.69)$$

For non-trivial solution of Equation 3.69, the following condition needs to be satisfied,

$$det(\begin{bmatrix} (k_1 + k_2) & -k_2 \\ -k_2 & k_2 \end{bmatrix} - \omega^2 \begin{bmatrix} m_1 & 0 \\ 0 & m_2 \end{bmatrix}) = 0$$

$$det(\begin{bmatrix} (k_1 + k_2) - \omega^2 m_1 & -k_2 \\ -k_2 & k_2 - \omega^2 m_2 \end{bmatrix}) = 0 \qquad (3.70)$$

Solution of Equation 3.70 gives two values ω^2 and thus two natural frequencies. Again, for each ω, A_1 and A_2 can be obtained such that the ratio A_1/A_2 is unique, though A_1 and A_2 are otherwise non-unique. This $\{A_1\ A_2\}^T$ obtained for each ω is the mode shape. In general, the non-uniqueness of the mode shapes is handled through normalization. Several normalization methods are used, for example, the first element of the mode shape vector is taken as unity and the other elements are modified accordingly to keep their ratio unchanged.

Though the method of obtaining natural frequencies and mode shapes has been explained using the example of a two-DOF system, the method can be directly extended for an arbitrary MDOF system. Let $\omega_1, \omega_2, \ldots, \omega_s, \ldots, \omega_n$ be the natural frequencies and $r^1, r^2, \ldots, r^s, \ldots, r^n$ be mode shapes of an MDOF with n dofs. Here,

$$r^1 = \begin{Bmatrix} r_{11} \\ r_{21} \\ \vdots \\ r_{s1} \\ \vdots \\ r_{n1} \end{Bmatrix}, \quad r^s = \begin{Bmatrix} r_{1s} \\ r_{2s} \\ \vdots \\ r_{ss} \\ \vdots \\ r_{ns} \end{Bmatrix}, \quad r^n = \begin{Bmatrix} r_{1n} \\ r_{2n} \\ \vdots \\ r_{sn} \\ \vdots \\ r_{nn} \end{Bmatrix} \qquad (3.71)$$

As mentioned earlier, the free vibration of the MDOF being the linear combination of the normal modes of vibration, it can be written as

$$\{u\} = \begin{Bmatrix} u_1 \\ u_2 \\ \vdots \\ u_s \\ \vdots \\ u_n \end{Bmatrix} = C_1 r^1 \cos(\omega_1 t - \varphi_1) + C_2 r^2 \cos(\omega_2 t - \varphi_2) + \ldots$$

$$+ C_s r^s \cos(\omega_s t - \varphi_s) + \ldots + C_n r^n \cos(\omega_n t - \varphi_n) \quad (3.72)$$

where the constants $C_1, C_2, \ldots, C_s, \ldots, C_n$ and the phase angles $\varphi_1, \varphi_2, \ldots, \varphi_s, \ldots, \varphi_n$ are obtained from initial conditions.

For all linear MDOF systems, the mass, $[M]$, and stiffness, $[K]$, matrices are symmetric and positive definite (positive semi-definite for system with rigid body modes). These properties of the matrices result in orthogonality of the mode shapes w.r.t. mass and stiffness matrices. This orthogonality can be expressed as

$$\{r^l\}^T[M]\{r^s\} = 0 \quad \{r^l\}^T[K]\{r^s\} = 0 \; \forall \, l \neq s \tag{3.73}$$

For mass normalized mode shapes, i.e.,

$$\{r^s\}^T[M]\{r^s\} = 1.0 \quad s = 1, \, 2, \ldots, n \tag{3.74}$$

Equations 3.73 can be rewritten as

$$[\phi]^T[M][\phi] = I \text{ and } [\phi]^T[K][\phi] = \begin{bmatrix} \ddots & & \\ & \omega_s^2 & \\ & & \ddots \end{bmatrix} \tag{3.75}$$

where $[\phi]$ is the modal matrix and is of the form

$$[\phi] = \begin{bmatrix} r_{11} & r_{12} & \cdots & r_{1s} & \cdots & r_{1n} \\ r_{21} & r_{22} & \cdots & r_{2s} & \cdots & r_{2n} \\ \vdots & \vdots & \cdots & \vdots & \cdots & \vdots \\ r_{s1} & r_{s2} & \cdots & r_{ss} & \cdots & r_{sn} \\ \vdots & \vdots & \cdots & \vdots & \cdots & \vdots \\ r_{n1} & r_{n2} & \cdots & r_{ns} & \cdots & r_{nn} \end{bmatrix} = \begin{bmatrix} r^1 & r^2 & \cdots & r^s & \cdots & r^n \end{bmatrix}$$

3.7 Modal analysis for forced vibration response of MDOF

Modal analysis is perhaps the most common analytical method used to obtain the response of an MDOF system. For a generalized undamped MDOF with n dofs, the EOM can be written as

$$[M]\{\ddot{u}\} + [K]\{u\} = \{F\} \tag{3.76}$$

where $[M]$ and $[K]$ are the $n \times n$ mass and stiffness matrices. $\{u\}$ and $\{F\}$ are vectors of length n containing the displacements and the applied forces. The underlying concept of modal analysis is to obtain a linear transformation matrix $[Q]$ as

$$\{u\} = [Q]\{q\} \tag{3.77}$$

Substituting, Equation 3.77 into Equation 3.76 and pre-multiplying both sides with $[Q]^T$ to maintain symmetry, we get

$$[Q]^T[M][Q]\{\ddot{q}\} + [Q]^T[K][Q]\{q\} = [Q]^T\{F\} \qquad (3.78)$$

Equation 3.78 will be solved for $\{q\}$ and $\{u\}$ using the inverse of Equation 3.77. The best $[Q]$ would be the one that will result in $[Q]^T[M][Q]$ and $[Q]^T[K][Q]$ to be diagonal. This will reduce the problem of solving a $n \times n$ coupled matrix equation given by Equation 3.76 to a problem of solving n decoupled equations.

Let the displacements $\{u\}$ be written as

$$\{u\} = r^1\eta_1(t) + r^2\eta_2(t) + \ldots + r^s\eta_s(t) + \ldots + r^n\eta_n(t) \qquad (3.79)$$

$$= [\phi]\left\{\begin{array}{c} \eta_1(t) \\ \eta_2(t) \\ \vdots \\ \eta_s(t) \\ \vdots \\ \eta_n(t) \end{array}\right\} = [\phi]\{\eta\} \qquad (3.80)$$

where $\{\eta\}$ is often referred to as a *generalized coordinate*. Substituting Equation 3.80 into Equation 3.78 and pre-multiplying both sides with $[\phi]^T$, we get

$$[\phi]^T[M][\phi]\{\ddot{\eta}\} + [\phi]^T[K][\phi]\{\eta\} = [\phi]^T\{F\} \qquad (3.81)$$

Using Equation 3.76 for mass normalized mode shapes/modal matrix, Equation 3.81 reduces to

$$\ddot{\eta}_s + \omega_s^2\eta_s = N_s \quad s = 1,\ 2,\ldots,n \qquad (3.82)$$

where N_s is the sth element of the transformed force vector $\{N\} = [\phi]^T\{F\}$. Equation 3.82 resembles the EOM for a SDOF with a frequency ω_s. Thus, Equation 3.82 can be solved applying the techniques adopted for SDOF and explained in the previous Sections. For example, for arbitrary loading, a convolution integral can be implemented. Once the solutions for all the elements of $\{\eta\}$ are obtained, $\{u\}$ can be derived using Equation 3.80. Similarly, the initial conditions given for $\{u\}$ can be converted to those for $\{\eta\}$ using the inverse of Equation 3.80.

The EOM of MDOF with damping is of the following form,

$$[M]\{\ddot{u}\} + [C]\{\dot{u}\} + [K]\{u\} = \{F\} \qquad (3.83)$$

Substituting Equation 3.80 into Equation 3.83 and pre-multiplying both sides with $[\phi]^T$ we get

$$[\phi]^T[M][\phi]\{\ddot{\eta}\} + [\phi]^T[C][\phi]\{\dot{\eta}\} + [\phi]^T[K][\phi]\{\eta\} = [\phi]^T\{F\} \qquad (3.84)$$

The transformed damping matrix $[\phi]^T[C][\phi]$ may not be diagonal and for such cases, the purpose of modal analysis to decouple the EOM is not met. For several cases, however, the damping matrix is given as

$$[C] = \alpha[M] + \beta[K] \tag{3.85}$$

The form of damping given by Equation 3.85 is referred to as proportional damping with α and β being constants. The other form of representing damping is also by providing the damping ratio ξ_s for the sth mode. The matrix $[\phi]^T[C][\phi]$ for this case can be directly written as a diagonal matrix with the sth diagonal term being $2\xi_s\omega_s$.

3.8 Response of MDOF system using wavelet transform

The implementation of wavelet transform to obtain the response of an MDOF is a direct extension of the method explained for SDOF in Section 3.5. The governing equation for an MDOF with m dofs is given by Equation 3.83 with the vector $\{u\}$ containing the displacements as mentioned earlier. Here, each of these displacements are approximated in time using Daubechies scaling function as given by Equation 3.45. Similarly, each element of the force vector needs to approximated as in Equation 3.46. The EOM given by Equation 3.83 following these approximations can be written as

$$[M][\Gamma^2]\{u_j\} + [C][\Gamma^1]\{u_j\} + [K]\{u_j\} = \{F_j\} \tag{3.86}$$
$$j = 0,\ 1,\ 2,\ldots,n-1$$
$$\tag{3.87}$$

Representing displacement of the lth dof as u_l, where, $l = 1,\ 2,\ldots,m$ and using Equation 3.65, $\{\ddot{u}_l\}$ can be written in terms of $\{u_l\}$. In addition, the matrix, $[\Gamma^1]$ can be written in the form,

$$[\Gamma^1] = \Phi\Pi\Phi^{-1} \tag{3.88}$$

where Φ is the eigenvector matrix of $[\Gamma^1]$ and Π is the diagonal matrix with diagonal terms being the eigenvalues which are represented as $-i\lambda_j$. $[\Gamma^2]$ can be written in similar form as

$$[\Gamma^2] = \Phi\Pi^2\Phi^{-1} \tag{3.89}$$

Thus, Equation 3.65 for this case can be written as

$$\tilde{\ddot{u}}_{lj} = -\lambda_j^2\tilde{u}_{lj} \quad \tilde{\dot{u}}_{lj} = -i\lambda_j\tilde{u}_{lj} \tag{3.90}$$

where

$$\tilde{u}_{lj} = \Phi^{-1}u_{lj} \quad \tilde{\ddot{u}}_{lj} = \Phi^{-1}\ddot{u}_{lj} \quad \tilde{\dot{u}}_{lj} = \Phi^{-1}\dot{u}_{lj} \tag{3.91}$$

Substituting Equations 3.90 and 3.91 in Equation 3.86, we get

$$-\lambda_j^2[M]\{\tilde{u}_j\} - i\lambda_j[C]\{\tilde{u}_j\} + [K]\{\tilde{u}_j\} = \{\tilde{F}_j\}$$
$$\left(-\lambda_j^2[M] - i\lambda_j[C] + [K]\right)\{\tilde{u}_j\} = \{\tilde{F}_j\} \qquad (3.92)$$
$$j = 0,\ 1,\ 2,\ldots,n-1$$

Equation 3.92 can be solved to obtain u_{lj}, $j = 0,\ 1,\ 2,\ldots,n-1$ and $l = 1,\ 2,\ldots,m$. The time domain response can be obtained by performing inverse transform. Similar to SDOF, the responses of an MDOF simulated using the present wavelet-based method match exactly with the responses obtained using a direct time integration scheme like Runge-Kutta 45.

Response of continuous systems can be obtained by modeling them using FE and then performing the time integration using the present method. FE model of the continuous system will result in a EOM similar to Equation 3.85 and can be solved using the same steps as used for the MDOF system. Continuous systems can also be solved in a different way using Daubechies scaling function approximation in time and space, but adopting a wave like solution. This is explained in later chapters on wavelet-based spectral finite element formulation. The WSFE formulation though has been used to simulate wave response of the continuous system under high frequency loading; such a model can also be used to derive the free and forced vibration responses of the continuous system.

Chapter 4

Wave Propagation: Spectral Analysis

In the earlier chapters, wavelet transform was used to study vibration problems. In the next part of the book, starting with this chapter, wavelet transform will be used to solve wave propagation problems. Before going into the details of wavelet-based solution procedure in the following chapters, this chapter gives an introduction to wave propagation, particularly spectral analysis of wave propagation. Such spectral analysis helps to study the frequency dependent wave characteristics and finally formulation of *Spectral Finite Element* (SFE) method. Here, the concepts of Fourier transform, mostly discrete Fourier transform (DFT) as explained in Chapter 2, are widely used. Solution of the problem in transformed wavelet domain being closely related to spectral analysis of wave propagation, this chapter is essential to comprehend the wavelet-based methodology explained later.

4.1 Spectrum and dispersion relations

Here, two important frequency dependent wave characteristics, namely, spectrum and dispersion relations, are obtained for a generalized system using DFT. These relations are the frequency variation of the wave parameters termed as *wavenumbers* and *wave speeds* respectively. These parameters are essential to understand the wave mechanics in a given waveguide and are also required for SFE formulation at a later stage. These parameters provide information like whether the wave mode is a propagating mode or a damping mode or a combination of these two (propagation as well as wave amplitude attenuation). Next, for a propagating mode, the nature of frequency variation of wavenumbers gives information whether the mode is *non-dispersive*, i.e., the wave retains its shape as it propagates or *dispersive* where the shape changes with propagation. In this section, these parameters are explained using the example of a generalized one-dimensional second- and fourth-order system.

The spectral analysis starts with the partial differential equation governing the waveguide. Considering a generalized second-order partial differential

given by

$$a\frac{\partial u}{\partial x^2} + b\frac{\partial u}{\partial x} = c\frac{\partial^2 u}{\partial t^2} \tag{4.1}$$

where a, b, and c are known constants depending on the material properties and geometry of the waveguide. $u(x,t)$ is the field variable to be solved for with x being the spatial dimension and t the temporal dimension. First, $u(x,t)$ is transformed to frequency domain using DFT as

$$u(x,t) = \sum_{n}^{N-1} \widehat{u}_n(x,\omega_n)e^{i\omega_n t} \tag{4.2}$$

where ω_n is the discrete circular frequency in *rad/sec* and N is the total number of frequency points used in the transformation. ω_n is related to the time window by

$$\omega_n = n\Delta\omega = \frac{n\omega_f}{N} = \frac{n}{N\Delta t} = \frac{n}{T}$$

where Δt is the time sampling rate and ω_f is the highest frequency captured by Δt. The frequency content of the load decides N and consideration of the *wrap around* and *aliasing* problem decides $\Delta\omega$. More details and associated problems are given in Reference [28]. Here, $\widehat{u}_n$ is the nth DFT coefficient and can also be referred to as the coefficient at frequency ω_n. $\widehat{u}_n$ varies only with x. Substituting Equation 4.2 into Equation 4.1, we get

$$a\frac{d^2\widehat{u}_n}{dx^2} + b\frac{d\widehat{u}_n}{dx} + c\omega_n^2\widehat{u}_n = 0 \quad n = 0,\,1,\ldots,N-1 \tag{4.3}$$

Thus, through DFT the governing PDE given by Equation 4.1 is reduced to N ODE varying only in x. Equations 4.3 being constant coefficient ODEs have a solution of the form $\widehat{u}_n = A_n e^{ik_n x}$, where A_n are the unknown constants which will be later derived from the boundary values and k_n is called the *wavenumbers* corresponding to the frequency ω_n. Substituting the above solution into Equations 4.3, we get the following characteristic equation to determine k_n,

$$\left(k_n^2 - i\frac{b}{a}k_n + \omega_n^2\frac{c}{a}\right)A_n = 0 \tag{4.4}$$

The subscript n is dropped hereafter for simplified notations. The above equation is quadratic in k and has two roots corresponding to the incident and reflected waves. If the wavenumbers are real, then the wave is called *propagating mode*. On the other hand, if the wavenumbers are complex, then the wave damps out as it propagates and hence is called *evanescent mode*. The set of the wavenumbers obtained by solving the characteristic Equation 4.4 is given as

$$k_{1,2} = i\frac{b}{2a} \pm \sqrt{-\frac{b^2}{4a^2} + \frac{c\omega^2}{a}} \tag{4.5}$$

Equation 4.5 is the generalized expression for the determination of the wavenumbers. Different wave behaviors are possible depending upon the values of the radical $\sqrt{c\omega^2/a - b^2/4a^2}$. As an example, for a case with $b = 0$, the wavenumbers are given as

$$k_{1,2} = \pm\omega\frac{c}{a} \qquad (4.6)$$

For such a case, the wavenumbers are real and hence the corresponding waves are propagating.

Next, two other important wave parameters, namely, *phase speed C_p* and *group speed C_g*, are introduced and briefly explained. They are defined as

$$C_p = \text{real}\left(\frac{\omega}{k}\right) \qquad C_g = \text{real}\left(\frac{d\omega}{dk}\right) \qquad (4.7)$$

Let us consider the previous example where wavenumbers vary linearly with frequency as given by Equation 4.6. Correspondingly, the wave speeds obtained using Equation 4.7 are as follows

$$C_p = C_g = \frac{a}{c} \qquad (4.8)$$

We find that both group and phase speeds are constant and equal. Hence, when wavenumbers vary linearly with frequency ω, the wave retains its shape as it propagates. Such waves are called *non-dispersive* waves. When wavenumbers have a non-linear variation with frequency, the phase and group speeds will not be constant but a function of frequency. As a result, each frequency component will travel with different speeds and the wave shape will not be preserved with wave propagation. Such waves are called *dispersive* waves.

Next, let us again consider the expression for wavenumber given by Equation 4.5 with all the constants being non-zero. This may result in a non-zero value of the radical giving complex or purely imaginary wavenumbers. Wavenumbers having significant real and imaginary parts imply that the corresponding wave will attenuate while it propagates. Here, the phase and wave speeds are given as

$$C_p = \text{real}\left(\frac{\omega}{\sqrt{c\omega^2/a - b^2/4a^2}}\right)$$

$$C_g = \text{real}\left(\frac{a\sqrt{c\omega^2/a - b^2/4a^2}}{c\omega}\right) \qquad (4.9)$$

Thus, it can be seen that these wave speeds C_p and C_g are not the same and hence the waves are dispersive in nature. There can also be a case where the wavenumbers are purely imaginary indicating that the waves will not propagate. Such a wave mode is a damping mode. The value of the radical however depends on frequency and there can be a frequency after which the wavenumbers transit from being purely imaginary to complex or real wavenumbers resulting in propagation of the wave mode. This transition frequency is called the *cut-off* frequency ω_c and can be derived by equating the radical to zero. The expression for the cut-off frequency for this second-order system is given as

$$\omega_c = \frac{b}{2\sqrt{ac}} \tag{4.10}$$

Once the wavenumbers are determined the solution of the transformed ODEs given by Equation 4.3 can be written as

$$\widehat{u}(x,\omega) = A_1 e^{-ik_1 x} + A_2 e^{-ik_2 x} \tag{4.11}$$

The unknown constants A_1 and A_2 can be evaluated in terms of the physical boundary conditions of the one-dimensional waveguide. This can be done in a formal manner using the *spectral finite element* technique which will be explained in details later. For $b = 0$ the above equation is of the form,

$$\widehat{u}(x,\omega) = A_1 e^{-ikx} + A_2 e^{ikx} \qquad k = \omega\sqrt{\frac{c}{a}} \tag{4.12}$$

where A_1 represent the incident wave coefficient while A_2 stands for the reflected wave coefficient.

Next, let us consider a fourth order system and study its wave behavior. Consider the following governing PDE

$$a\frac{\partial^4 w}{\partial x^4} + bw + c\frac{\partial^2 w}{\partial t^2} = 0 \tag{4.13}$$

where $w(x,t)$ is the field variable and a, b, c are arbitrary known constants depending on the material and geometric properties of the waveguide as in the case of the second-order system. The above equation is similar to the equation of motion of a Euler-Bernoulli beam on elastic foundation. The DFT of $w(x,t)$ can be written in a similar form as Equation 4.2,

$$w(x,t) = \sum_{0}^{N-1} \widehat{w}_n(x,\omega_n)e^{i\omega_n t} \tag{4.14}$$

Substituting Equation 4.14 into the governing PDE given by Equation 4.13 we get the reduced ODEs as

$$a\frac{d^4 \widehat{w}_n}{dx^4} - (c\omega_n^2 - b)\widehat{w}_n = 0 \qquad n = 0,\ 1,\ldots,N-1 \tag{4.15}$$

The above ODEs have constant coefficients and hence the solution will be of the form $\widehat{w}_n = A_n e^{ik_n x}$. Substituting this solution into Equation 4.15 we get the characteristic equation for solution of the wavenumbers. Again the subscript n is dropped hereafter for simplified notations and all the following equations have to be derived for n varying from 0 to $N-1$. The characteristic equation is of the form

$$k^4 - \left(\frac{c}{a}\omega^2 - \frac{b}{a}\right) = 0 \qquad (4.16)$$

This is a fourth order equation and will give two sets of wavenumbers. The type of wave is dependent upon the numerical value of $\frac{c\omega^2}{a} - \frac{b}{a}$. For $\frac{c\omega^2}{a} > \frac{b}{a}$, the solution of Equation 4.16 will give the following wavenumbers,

$$k_{1,2} = \pm\beta \qquad k_{3,4} = \pm i\beta \qquad (4.17)$$

where $\beta = \left|\left(\frac{c}{a}\omega^2 - \frac{b}{a}\right)^{\frac{1}{4}}\right|$. In the above equation k_1 and k_2 represent the propagating wave modes while k_3 and k_4 are the damping or evanescent modes. From the above equations, we find that the wavenumbers are non-linear functions of the frequency and hence the corresponding waves are expected to be highly dispersive in nature. Also, using the above expression we can find the phase and group speeds for the propagating mode from Equation 4.7.

Next, consider the case when $\frac{c\omega^2}{a} < \frac{b}{a}$. For such conditions, the wavenumbers are given by

$$k_{1,2} = \pm\frac{1}{\sqrt{2}}(1+i)\beta \qquad k_{3,4} = \pm\frac{1}{\sqrt{2}}(-1+i)\beta \qquad (4.18)$$

From the above equation, we see that the change of sign of $\left(\frac{c\omega^2}{a} - \frac{b}{a}\right)$ has completely changed the wave behavior. Now, all the wavenumbers have both real and imaginary parts. Hence all the wave modes are propagating as well as attenuating. The initial evanescent mode also becomes a propagating mode after the cut-off frequency ω_c. The expression for the cut-off frequency obtained by equating $\left(\frac{c\omega^2}{a} - \frac{b}{a}\right)$ to zero is

$$\omega_c = \sqrt{\frac{b}{c}} \qquad (4.19)$$

Again, if $b = 0$, the cut-off frequency vanishes and the wave behavior is similar to the first case, i.e., it will have propagating and damping modes. In all cases, however, the waves will be highly dispersive in nature.

The solution of the fourth-order governing equation 4.15 can be written as

$$\hat{w}(x,\omega) = A_1 e^{-i\beta x} + B_1 e^{-\beta x} + A_2 e^{i\beta x} + B_2 e^{\beta x} \qquad (4.20)$$

As in the previous case, A_1, B_1 are the incident wave coefficients and A_2, B_2 are the reflected wave coefficients. These unknown constants can be determined in terms of the physical boundary conditions of the beam.

From the above discussion, we see that the spectral analysis gives an insight into the wave mechanics of a system defined by its governing differential equation. Though spectral analysis can be done similarly using wavelet transform, it is not as straightforward as Fourier transform-based analysis. This is because the wavelet basis functions are bounded both in time and frequency unlike the basis for Fourier transform which is unbounded in time. This has been explained in greater detail with examples in the other chapters.

4.2   Computations of wavenumbers and wave amplitudes

In the last section, the parameters, wavenumber and wave speed were explained with the examples of generalized second-order and fourth order partial differential wave equations. The wavenumbers k were obtained as a function of frequency ω by solving second and fourth order polynomial equations respectively. The computation of wavenumbers is however not so straightforward for structures with higher complexities. For one-dimensional structures such cases arise when the governing equation is a set of coupled PDEs and a couple of such examples are Timoshenko beam and other higher order beams. In a Timoshenko beam, the governing equations consist of two coupled PDEs with transverse and shear displacements as the variables. Another common example of structure having a set of coupled PDEs as governing equations is a composite beam with asymmetric ply lay-up resulting in elastic coupling. In addition to the different one-dimensional structures, computation of wavenumbers for two-dimensional structures is also difficult primarily because the wavenumbers here are a function of both frequency and wavenumber in the other direction.

In order to handle such problems, generalized and computationally implementable methods have been proposed [36] to calculate the wavenumbers and associated wave amplitude. The two different approaches to solve the problem are based on *singular value decomposition* (SVD) and *polynomial eigenvalue problem* (PEP) methods. The methods are described briefly in the following Subsections using the example of a Timoshenko beam. Though the methods are explained using the example of an isotropic Timoshenko beam, they can be implemented for any arbitrary structure in a similar way.

The governing differential equations for Timoshenko beam are given as

$$GA\left[\frac{\partial^2 v}{\partial x^2} - \frac{\partial \phi}{\partial x}\right] = \rho A \frac{\partial^2 v}{\partial t^2} \tag{4.21}$$

$$EI\frac{\partial^2 \phi}{\partial x^2} + GA\left[\frac{\partial v}{\partial x} - \phi\right] = \rho I \frac{\partial^2 \phi}{\partial t^2} \tag{4.22}$$

where v and ϕ are the transverse and shear, displacements respectively. E, G are the Young's modulus and shear modulus, ρ is the mass density. A and I are the cross-sectional area and area moment of inertia. Similar to Equations 4.2

and 4.14, the displacements $v(x,t)$ and $\phi(x,t)$ are transformed to frequency domain using DFT as follows,

$$v(x,t) = \sum_{n=0}^{N-1} \widehat{v}_n(x,\omega_n)e^{i\omega_n t} \quad \text{and} \quad \phi(x,t) = \sum_{n=0}^{N-1} \widehat{\phi}_n(x,\omega_n)e^{i\omega_n t} \qquad (4.23)$$

Substituting the above approximation into the Equations 4.21 and 4.22, and dropping the subscript n for simplified notations, we get

$$GA\left[\frac{d^2\widehat{v}}{dx^2} - \frac{d\phi}{dx}\right] = -\omega^2\rho A\widehat{v} \qquad (4.24)$$

$$EI\frac{d^2\widehat{\phi}}{dx^2} + GA\left[\frac{d\widehat{v}}{dx} - \widehat{\phi}\right] = -\omega^2\rho I\widehat{\phi}. \qquad (4.25)$$

Assuming the solution of the above equations as $\widehat{v} = C_v e^{ikx}$ and $\widehat{\phi} = C_\phi e^{ikx}$ and substituting them into those equations we get the following polynomial equations,

$$GA\left[-k^2 C_v - ikC_\phi\right] = -\omega^2\rho AC_v \qquad (4.26)$$

$$-k^2 EIC_\phi + GA\left[ikC_v - C_\phi\right] = -\omega^2\rho IC_\phi \qquad (4.27)$$

As said before, the formal ways of solving the above equations to obtain the wavenumbers k and C_v, C_ϕ are methods based on SVD and PEP as explained in the two following Subsections.

4.2.1 SVD-based technique

In this method, Equations 4.26 and 4.27 can be written in the matrix form $\mathbf{W}(k)\mathbf{C} = 0$, where

$$\mathbf{W}(k) = \begin{bmatrix} -k^2 GA + \omega^2\rho A & -ikGA \\ ikGA & -k^2 EI - GA + \omega^2\rho I \end{bmatrix} \quad \text{and} \quad \mathbf{C} = \begin{Bmatrix} C_v \\ C_\phi \end{Bmatrix} \qquad (4.28)$$

Here, it is noted that the ks are the latent roots of the above equation, which satisfy the conditions $det(\mathbf{W}(k)) = 0$ [44]. Further for each solution of k, there is at least one non-trivial solution for $\mathbf{C}$ which is known as the latent eigenvector.

To find the latent roots, the determinant is expanded in a polynomial of k, $p(k)$, and solved by the *companion matrix* method. In this method, the companion matrix $L(p)$, corresponding to $p(k)$ is formed, which is defined as

$$L(p) = \begin{bmatrix} 0 & 1 & 0 & \cdots & 0 \\ 0 & 0 & 1 & \cdots & 0 \\ \vdots & \vdots & \vdots & \ddots & \vdots \\ 0 & 0 & \cdots & 0 & 1 \\ -\alpha_m & -\alpha_{m-1} & \cdots & -\alpha_2 & -\alpha_1 \end{bmatrix} \qquad (4.29)$$

where $p(k)$ is given by

$$p(k) = k^m + \alpha_1 k^{m-1} + \ldots + \alpha_m \qquad (4.30)$$

One of the many important properties of the companion matrix is that the characteristic polynomial of $L(p)$ is $p(k)$ itself [45]. For the Timoshenko beam, the companion matrix $L(p)$ derived from $\mathbf{W}(k)$ in Equation 4.28 is of the form

$$L(p) = \begin{bmatrix} 0 & 1 & 0 & 0 \\ 0 & 0 & 1 & 0 \\ 0 & 0 & 0 & 1 \\ \omega^2 \frac{\rho A}{EI} - \omega^4 \frac{\rho A}{GA} \frac{\rho I}{EI} & 0 & \omega^2 \left(\frac{\rho I}{EI} + \frac{\rho A}{GA} \right) & 0 \end{bmatrix} \qquad (4.31)$$

The eigenvalues of $L(p)$ are the roots of $p(k)$, which are obtained using standard techniques.

Once the eigenvalues are obtained they are used to obtain the eigenvectors. The eigenvectors are the elements of the null space of $\mathbf{W}(k)$ and the eigenvalues make this null space non-trivial by rendering $\mathbf{W}(k)$ singular. Hence, computation of the eigenvectors is equivalent to computation of the null space of a matrix. For this purpose, the SVD method is most effective. $\mathbf{U}$, $\mathbf{V}$, and the diagonal matrix $\mathbf{S}$ as $\mathbf{A} = \mathbf{USV^H}$, where the superscript $\mathbf{H}$ denotes Hermitian conjugate. $\mathbf{S}$ is the singular value matrix. For singular matrices, one or more of the singular values will be zero and the required property of the unitary matrix $\mathbf{V}$ is that the columns of $\mathbf{V}$ corresponding to zero singular values are the elements of the null space of $\mathbf{A}$. Again several standard techniques are available to perform SVD.

4.2.2   PEP-based technique

In this approach, equations similar to Equations 4.26 and 4.27 are written as PEP in k as

$$\mathbf{A_0} k^m + \mathbf{A_1} k^{m-1} + \ldots + \mathbf{A_m} = 0 \qquad (4.32)$$

where $\mathbf{A_i}$, $i = 0$ to m are the $p \times p$ matrices, where p is the number of independent variables in the governing equations. Thus, the PEP is of the order $p \times m$. For Timoshenko beam, $p = 2$ and $m = 2$. The PEP for Equations 4.26 and 4.27 is given as

$$\mathbf{A_0} k^2 + \mathbf{A_1} k + \mathbf{A_2} = 0 \qquad (4.33)$$

where

$$\mathbf{A_0} = \begin{bmatrix} -GA & 0 \\ 0 & -EI \end{bmatrix} \quad \mathbf{A_1} = \begin{bmatrix} 0 & -iGA \\ iGA & 0 \end{bmatrix} \quad \mathbf{A_2} = \begin{bmatrix} \omega^2 \rho A & 0 \\ 0 & -GA + \omega^2 \rho I \end{bmatrix}$$

The wavenumbers k are obtained as eigenvalues of the PEP while the wave amplitudes are the eigenvectors. The PEP yields $p \times m$ eigenvalues and

m eigenvectors. If both $\mathbf{A_0}$ and $\mathbf{A_m}$ are singular the problem is potentially ill-posed. Theoretically the solution might not exist or might not be unique. Computationally, the computed solution may be inaccurate. If one, but not both $\mathbf{A_0}$ and $\mathbf{A_m}$ is singular, the problem is well posed, but some of the eigenvalues may be zero or infinite and caution should be maintained in rejecting those roots.

There are advantages and disadvantages of both the methods. In the first method, the determinant of the wave matrix needs to be formed, which for large p is difficult to obtain. In this case, resorting to the second method is advantageous as it doesn't require to obtain the long expression for α_i in Equation 4.30. In the second method, however, there is no control over the eigenvalues, as we might be interested sometimes in separating the wavenumbers for different modes. In such cases, the first method is more suited.

4.3 Spectral finite element (SFE) method

The spectral finite element method in brief is a FE method in the frequency domain. There are, however, some essential differences in implementation of these methods. The main differences are highlighted in the following paragraph.

FE formulation assumes a polynomial function for the displacement variation within the element. These assumed polynomials are forced to satisfy the weak form of the governing differential equation, which yields the stiffness and mass matrices. These elemental matrices are assembled to obtain the global stiffness and mass matrices. The assembly process ensures equilibrium of forces between the adjacent elements. Finally, the continuum is represented by a discretized model governed by the matrix equation,

$$[M]\{\ddot{u}\} + [K]\{u\} = \{F(t)\} \qquad (4.34)$$

where $[M]$ and $[K]$ are the global mass and stiffness matrices. $\{\ddot{u}\}$ and $\{u\}$ are the acceleration and displacement vectors. Equation 4.34 represents the FE model of an undamped structure and additional terms are introduced to incorporate damping.

Equation 4.34 is an ODE in time. One of the conventional methods of solving such equation, is the mode superposition method, which, however, is computationally prohibitive for wave propagation analysis. The methods that can be implemented are the time marching schemes [3] which can either be the *explicit* or *implicit* methods. For the transient dynamic problem of wave propagation the explicit methods are normally preferred. In the time marching scheme, the solution process takes place over a small time step Δt and is repeated for N time steps where $N\Delta t$ is the total time window.

The solution of the dynamic equation will give displacement, velocity, and acceleration histories. The solution time being directly proportional to the number of degrees of freedom, the computational time of such a scheme is very high for wave propagation problems.

SFE method on the other hand uses in most cases the exact solution to the wave equation as its interpolating function. For example, the interpolating functions given by Equations 4.12 and 4.20 for the second and fourth-order systems described in the previous section are exact solutions of the transformed equations. In SFE, however, these exact interpolating functions are complex exponential in place of polynomials used in FE. These exponential interpolation functions give the dynamic stiffness matrices after imposition of boundary conditions at the nodes. The dynamic stiffness matrix relates the transformed nodal displacements with the transformed nodal forces analogous to the FE method. The basic steps involved in Fourier transform-based SFE are as follows. First, the given forcing function is transformed to the frequency domain using the forward FFT. For this, we need to choose the time sampling rate and number of FFT points to decide on the analysis time window. The length of the chosen time window should be more than that required to avoid the *wrap around* problem [28]. The FFT output will yield the frequency, and the real and imaginary parts of the forcing function, which are stored separately. Next, the elemental stiffness matrix is generated, assembled, and solved over the frequency loop. The procedures of assembling and solution of the dynamic stiffness matrix are exactly similar to that adopted in the conventional FE method. The main difference is however that in SFE these processes are carried at every frequency sampling point. This does not pose a major computational hurdle since the problem sizes are many orders smaller than conventional FE. The solution process is first performed for a unit impulse, which directly yields the *Frequency Response Function* (FRF). The FRF is then convolved with the load to get the required output in the frequency domain. The output is then transformed to the time domain using the inverse FFT.

As mentioned earlier, SFE method is especially tailored for solving wave propagation problems and possesses several advantages over conventional FE method in this respect. The primary advantage is computational efficiency. In absence of discontinuity, a single SFE is sufficient to accurately model one- or two-dimensional waveguides. This reduces the problem size by orders of magnitude as compared to FE method. The SFE method however follows the procedures of FE formulation and thus can model structures of higher complexities in contrast to the other analytical or semi-analytical simulation techniques. Next, SFE method allows simultaneous time and frequency domain analysis unlike FE method. For wave propagation problems, frequency domain analysis is equally important as time domain analysis. Several important wave characteristics can be extracted only from the frequency domain. Finally, the transfer function relating the output (displacement, velocity, etc.) to the input (force, stress, etc.) can be derived from the SFE formulation and thus can be used efficiently for solving inverse problems. There is much literature implementing SFE for the force identification problem [84, 27, 85, 29, 62].

Though SFE possess several properties in favor of their efficient use for simulation of wave propagation in structural waveguides, the conventional Fourier transform-based SFE has some serious limitations which restrict their wider use. These limitations arise from the inability of FSFE to model finite waveguides, both one-dimensional and two-dimensional. As explained earlier, this was primarily because the basis function used in FFT has global support from $-\infty$ to ∞. It is also due to induced periodicity both in time and frequency domains. The Daubechies wavelet basis functions having localized support was foreseen to have the capability to model finite dimension waveguides when used as basis function for SFE formulation. This led to the development of wavelet-based SFE (WSFE) [62, 63]. The WSFE method can accurately simulate wave propagation in finite dimension waveguides while retaining all the other advantages of SFE. The following chapters of this book explain in detail the formulation of WSFE for one- and two-dimensional waveguides and also application of the developed WSFE models for analysis of structures with higher complexities like carbon nanotubes and their composites.

A considerable portion of the WSFE formulation is similar to that of the FSFE formulation. Thus in the following two sections, the FSFE formulation of an example of one-dimensional and two-dimensional waveguides is explained in brief for the readers to get acquainted with some basic steps of SFE formulations, which remain invariant of the nature of basis functions used, i.e., wavelet or Fourier transforms. Greater details of FSFE can be obtained in the books dedicated to FSFE [28, 36]

4.4 FSFE formulation of Timoshenko beam

In this section, FSFE formulation for one-dimensional waveguides is briefly illustrated using the example of Timoshenko beam. The governing differential equations for the beam are given by Equations 4.21 and 4.22. Following Fourier transform in time as explained in the previous sections, the partial differential wave equations are reduced into a set of ODEs given by Equations 4.24 and 4.25. These equations are solved exactly to obtain the exact shape functions with unknown constants. These unknown constants are solved in terms of boundary values and are used to formulate the elemental dynamic stiffness matrix relating the transformed nodal displacements with transformed nodal forces.

The degrees of freedom associated with the Timoshenko beam element formulation are shown in Figure 4.1. The element has two degrees of freedom per node, namely, $\widehat{v}$ and $\widehat{\phi}$. Equations 4.24 and 4.25 need to be solved for $\widehat{v}$

FIGURE 4.1: Timoshenko beam element with nodal displacements and nodal forces

and $\widehat{\phi}$, and the actual solutions $w(x,t)$ and $\phi(x,t)$ are obtained using inverse FFT. The exact shape functions obtained by solving Equations 4.24 and 4.25 are of the form

$$\left\{ \begin{array}{c} \widehat{w}(x) \\ \widehat{\phi}(x) \end{array} \right\} = [\mathbf{R}][\mathbf{\Theta}]\{C\} \tag{4.35}$$

where $[\mathbf{\Theta}]$ is a diagonal matrix with diagonal terms $[e^{-ik_1 x}, e^{-ik_1(L-x)}, e^{-ik_2 x}, e^{-ik_2(L-x)}]$. $[\mathbf{R}]$ is the 2×4 amplitude ratio matrix obtained by solving the PEP of Equation 4.33, in addition to the wavenumbers $\pm k_1$ and $\pm k_2$. $[\mathbf{R}]$ is the eigenvector matrix while the wavenumbers $\pm k_1$ and $\pm k_2$ are the eigenvalues of the above mentioned PEP. The $[\mathbf{R}]$ has the following form,

$$[\mathbf{R}] = \begin{bmatrix} R_{11} & R_{12} & R_{13} & R_{14} \\ R_{21} & R_{22} & R_{23} & R_{24} \end{bmatrix} \tag{4.36}$$

Here, $\{C\} = \{C_1\ C_2\ C_3\ C_4\}$ are the unknown constants. Hence, the expression of each transformed displacement can be written as

$$\widehat{w}(x) = C_1 R_{11} e^{-ik_1 x} + C_2 R_{12} e^{-ik_1(L-x)}$$
$$+ C_3 R_{13} e^{-ik_2 x} + C_4 R_{14} e^{-ik_2(L-x)} \tag{4.37}$$
$$\widehat{\phi}(x) = C_1 R_{21} e^{-ik_1 x} + C_2 R_{22} e^{-ik_1(L-x)}$$
$$+ C_3 R_{23} e^{-ik_2 x} + C_4 R_{24} e^{-ik_2(L-x)} \tag{4.38}$$

The above equations are the expanded form of the matrix Equation 4.35. Imposing the boundary conditions at the two nodes as shown in Figure 4.1, the unknown constants $\{C\}$ can be obtained in terms of the transformed nodal displacements and forces. The transformed nodal displacements are represented by $\{\widehat{\mathbf{u}}^{\mathbf{e}}\}$, where $\{\widehat{\mathbf{u}}^{\mathbf{e}}\} = \{\widehat{w}_1\ \widehat{\phi}_1\ \widehat{w}_2\ \widehat{\phi}_2\}$ and $\widehat{v}_1 \equiv \widehat{w}(0)$, $\widehat{\phi}_1 \equiv \widehat{\phi}(0)$, $\widehat{v}_2 \equiv \widehat{w}(L)$, $\widehat{\phi}_2 \equiv \widehat{\phi}(L)$. Substituting these into Equations 4.37 and 4.38 and writing them in a matrix form, we get

$$\left\{ \begin{array}{c} \widehat{w}_1 \\ \widehat{\phi}_1 \end{array} \right\} = [\mathbf{R}][\mathbf{\Theta}_1]\{C\} = [\mathbf{T_{11}}]\{C\}$$

$$\left\{ \begin{array}{c} \widehat{w}_2 \\ \widehat{\phi}_2 \end{array} \right\} = [\mathbf{R}][\mathbf{\Theta}_2]\{C\} = [\mathbf{T_{12}}]\{C\}$$

$$\{\widehat{\mathbf{u}}^{\mathbf{e}}\} = \begin{Bmatrix} \widehat{w}_1 \\ \widehat{\phi}_1 \\ \widehat{w}_2 \\ \widehat{\phi}_2 \end{Bmatrix} = \begin{bmatrix} \mathbf{T_{11}} \\ \mathbf{T_{12}} \end{bmatrix} \{C\} = [\mathbf{T_1}]\{C\} \tag{4.39}$$

where $[\mathbf{\Theta_1}]$ and $[\mathbf{\Theta_2}]$ are obtained by substituting $x = 0$ and $x = L$ in the matrix $[\mathbf{\Theta}]$. Next, the force boundary conditions need to be enforced to obtain the unknown constants $\{C\}$ in terms of the transformed nodal forces. To obtain this relation the following steps are followed. The force boundary conditions associated with the governing differential equation of a Timoshenko beam are given as

$$GA\left[\frac{\partial w}{\partial x} - \phi\right] = V \tag{4.40}$$

$$EI\frac{\partial \phi}{\partial x} = M \tag{4.41}$$

where $V(x,t)$ and $M(x,t)$ are the shear force and moment. Transforming Equations 4.40 and 4.41 through FFT on both sides, similar to the governing differential equations, we get a set of reduced ODEs as

$$GA\left[\frac{d\widehat{w}}{dx} - \widehat{\phi}\right] = \widehat{V} \tag{4.42}$$

$$EI\frac{d\widehat{\phi}}{dx} = \widehat{M} \tag{4.43}$$

where $\widehat{V}(x)$ and $\widehat{M}(x)$ are the transformed shear force and moment. Substituting Equations 4.37 and 4.38 into Equations 4.42 and 4.43, we get the expression of $\widehat{V}(x)$ and $\widehat{M}(x)$ as

$$\widehat{V}(x) = C_1(-ik_1R_{11} - R_{21})e^{-ik_1x} + C_2(ik_1R_{12} - R_{22})e^{-ik_1(L-x)} +$$
$$C_3(-ik_2R_{13} - R_{23})e^{-ik_2x} + C_4(ik_2R_{14} - R_{24})e^{-ik_2(L-x)} \tag{4.44}$$

$$\widehat{M}(x) = C_1(-ik_1R_{21})e^{-ik_1x} + C_2(ik_1R_{22})e^{-ik_1(L-x)} +$$
$$C_3(-ik_2R_{23})e^{-ik_2x} + C_4(ik_1R_{24})e^{-ik_2(L-x)} \tag{4.45}$$

Equations 4.44 and 4.45 can be written in a matrix form as

$$\begin{Bmatrix} \widehat{V}(x) \\ \widehat{M}(x) \end{Bmatrix} = [\mathbf{R'}][\mathbf{\Theta}]\{C\} \tag{4.46}$$

where $[\mathbf{R'}]$ has the following form,

$$[\mathbf{R'}] = \begin{bmatrix} (-ik_1R_{11} - R_{21}) & (ik_1R_{12} - R_{22}) & (-ik_2R_{13} - R_{23}) & (ik_2R_{14} - R_{24}) \\ -ik_1R_{21} & ik_1R_{22} & -ik_2R_{23} & ik_2R_{24} \end{bmatrix} \tag{4.47}$$

Let the transformed nodal forces be represented by $\{\widehat{\mathbf{F}}^e\}$, where $\{\widehat{\mathbf{F}}^e\} = \{\widehat{V}_1\ \widehat{M}_1\ \widehat{V}_2\ \widehat{M}_2\}$ and $\widehat{V}_1 \equiv \widehat{V}(0)$, $\widehat{M}_1 \equiv \widehat{M}(0)$, $\widehat{V}_2 \equiv \widehat{V}(L)$, and $\widehat{M}_2 \equiv \widehat{M}(L)$. Substituting these boundary conditions into Equations 4.44 and 4.45 and writing them in matrix form, we get

$$\left\{ \begin{array}{c} \widehat{V}_1 \\ \widehat{M}_1 \end{array} \right\} = [\mathbf{R}'][\mathbf{\Theta_1}]\{C\} = [\mathbf{T_{21}}]\{C\}$$

$$\left\{ \begin{array}{c} \widehat{V}_2 \\ \widehat{M}_2 \end{array} \right\} = [\mathbf{R}'][\mathbf{\Theta_2}]\{C\} = [\mathbf{T_{22}}]\{C\}$$

$$\{\widehat{\mathbf{F}}^e\} = \left\{ \begin{array}{c} \widehat{V}_1 \\ \widehat{M}_1 \\ \widehat{V}_2 \\ \widehat{M}_2 \end{array} \right\} = \left[\begin{array}{c} \mathbf{T_{21}} \\ \mathbf{T_{22}} \end{array} \right] \{C\} = [\mathbf{T_2}]\{C\} \tag{4.48}$$

Combining Equations 4.39 and 4.48, we can relate the transformed nodal displacements with transformed nodal forces as

$$\{\widehat{\mathbf{F}}^e\} = [\mathbf{T_2}][\mathbf{T_1}]^{-1}\{\widehat{\mathbf{u}}^e\} = [\widehat{\mathbf{K}}^e]\{\widehat{\mathbf{u}}^e\} \tag{4.49}$$

where $[\widehat{\mathbf{K}}^e]$ is the elemental dynamic stiffness matrix. The elemental stiffness matrix can be then either solved to obtain $\{\widehat{\mathbf{u}}^e\}$ for given forcing and boundary conditions similar to conventional FE. In addition, the elemental matrix can be assembled to obtain the global dynamic stiffness matrix. The process of assembly is also exactly similar to that in conventional FE.

Once the transformed nodal displacements $\{\widehat{\mathbf{u}}^e\}$ are known from Equation 4.49, they can be substituted into Equation 4.39 to derive the values of $\{C\}$. Next, these values of $\{C\}$ can be used in Equation 4.35 to obtain the displacement at any arbitrary spatial position x.

4.5   FSFE formulation of isotropic plate under in-plane loading

In this section, similar to the previous section, FSFE formulation for a two-dimensional waveguide is explained through the example of an isotropic plate experiencing in-plane deformation only. The governing differential equations for such structure are given as

$$(\lambda + 2\mu)\frac{\partial^2 u}{\partial x^2} + (\lambda + \mu)\frac{\partial^2 v}{\partial x \partial y} + \mu\frac{\partial^2 u}{\partial y^2} = \rho\ddot{u} \tag{4.50}$$

$$\nu\frac{\partial^2 v}{\partial x^2} + (\lambda + \mu)\frac{\partial^2 u}{\partial x \partial y} + (\lambda + 2\mu)\frac{\partial^2 v}{\partial y^2} = \rho\ddot{v} \tag{4.51}$$

Here, $u(x, y, t)$ and $v(x, y, t)$ are the displacements in X and Y directions. λ and μ are elastic constants conventionally defined as $\lambda = \frac{\nu E}{(1+\nu)(1-2\nu)}$ and $\mu = \frac{E}{2(1+\nu)}$ The solution of these displacement fields are assumed to be of the form [36]

$$u(x, y, t) = \sum_{n=0}^{N-1} \sum_{m=0}^{M-1} \widehat{u}(x, \xi_m, \omega_n) e^{i\xi_m y} e^{i\omega_n t} \tag{4.52}$$

$$v(x, y, t) = \sum_{n=0}^{N-1} \sum_{m=0}^{M-1} \widehat{v}(x, \xi_m, \omega_n) e^{i\xi_m y} e^{i\omega_n t} \tag{4.53}$$

where ξ_m is the discrete lateral wavenumbers. The discrete values of ξ_m depend on the Y window length and number of lateral mode shapes (M) chosen.

Substituting Equations 4.52 and 4.53 in the governing equations 4.50 and 4.51, we get a set of $N \times M$ ODEs as follows,

$$(\lambda + 2\mu)\frac{d^2\widehat{u}}{dx^2} + i\xi_m \frac{d\widehat{v}}{dx} - \xi_m^2 \mu\widehat{u} = -\omega_n^2 \rho\widehat{u} \tag{4.54}$$

$$\nu\frac{d^2\widehat{v}}{dx^2} + i\xi_m(\lambda + \nu)\frac{d\widehat{u}}{dx} - \xi_m^2(\lambda + 2\mu)\widehat{v} = -\omega_n^2 \rho\widehat{v} \tag{4.55}$$

The above equations need to be solved for $\widehat{u}$ and $\widehat{v}$. The procedure is similar to that followed for FSFE formulation of Timoshenko beam explained in Sections 4.2 and 4.4. Assuming the solution of Equations 4.54 and 4.55 as $\widehat{u} = C_u e^{ikx}$ and $\widehat{v} = C_v e^{ikx}$ and substituting them in the equations gives

$$-k^2(\lambda + 2\mu)C_u - k\xi_m(\lambda + \mu)C_v - \xi_m^2 \mu C_u = -\omega_n^2 \rho C_u \tag{4.56}$$

$$-k^2\nu C_v - k\xi_m(\lambda + \mu)C_u - \xi_m^2 C_v = -\omega_n^2 \rho C_v \tag{4.57}$$

The subscripts n and m are dropped hereafter for simplified notations; however, all the following equations hold for $n = 0$ to $N - 1$ and $m = 0$ to $M - 1$. The above equations can be solved to obtain two sets of conjugate wavenumbers k and 2×4 wave amplitude ratio matrix $[\mathbf{R}]$ either through SVD method or through solution of PEP matrix as explained in Section 4.2. When posed as PEP, Equations 4.56 and 4.57 are of similar form as Equation 4.33, where

$$\mathbf{A_0} = \begin{bmatrix} -(\lambda + 2\mu) & 0 \\ 0 & -\mu \end{bmatrix} \quad \mathbf{A_1} = \begin{bmatrix} 0 & -\xi(\lambda + \mu) \\ -\xi(\lambda + \mu) & 0 \end{bmatrix}$$

$$\mathbf{A_2} = \begin{bmatrix} (\omega^2\rho - \xi^2\mu) & 0 \\ 0 & (\omega^2\rho - \xi^2(\lambda + 2\mu)) \end{bmatrix} \tag{4.58}$$

The exact shape functions for this two-dimensional FSFE formulation obtained by solving the PEP for given values of ξ and ω are of the following form,

$$\left\{ \begin{array}{c} \widehat{u}(x) \\ \widehat{v}(x) \end{array} \right\} = [\mathbf{R}][\Theta]\{C\} \tag{4.59}$$

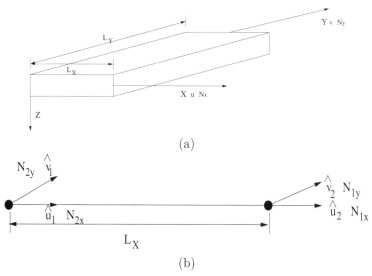

FIGURE 4.2: (a) Plate element; (b) nodal displacements and forces

Equation 4.59 is similar to Equation 4.35 and $[\mathbf{R}]$, $[\mathbf{\Theta}]$, and $\{C\}$ in Equation 4.59 have similar form as in Equation 4.35, but certainly different values. The formulation of FSFE for two-dimensional plate with in-plane deformation hereafter is similar to the FSFE formulation for the one-dimensional Timoshenko beam to a large extent.

The degrees of freedom associated with FSFE of this two-dimensional wave guide are shown in Figure 4.2(a) and (b). The element has two nodes, one at $x = 0$ and the other at $x = L_X$. Each node has two degrees of freedom, namely, transformed displacements $\widehat{u}$ and $\widehat{v}$ in X and Y directions respectively. The load can be applied along any line parallel to Y-axis

Next, the unknown constants $\{C\}$ in Equation 4.59 can be obtained in terms of the transformed nodal displacements $\{\{\widehat{\mathbf{F}}^{\mathbf{e}}\}\}$ through the matrix relation given by

$$\{\widehat{\mathbf{u}}^{\mathbf{e}}\} = \left\{ \begin{array}{c} \widehat{u}_1 \\ \widehat{v}_1 \\ \widehat{u}_2 \\ \widehat{v}_2 \end{array} \right\} = [\mathbf{T_1}]\{C\} \tag{4.60}$$

The derivation of Equation 4.60 from Equation 4.59 can be done following the steps similar to that given by Equations 4.37 to 4.39 in Section 4.4.

Obtaining the unknown constant $\{C\}$ in terms of the transformed nodal displacements, the only task left in formulating FSFE is to relate these constants to the transformed nodal forces. The process to be followed is similar to that adopted for one-dimensional waveguide formulation in Section 4.4. The force boundary conditions used for the two-dimensional waveguides are however different and are given as

$$N_x = (\lambda + 2\mu)\frac{\partial u}{\partial x} + \lambda\frac{\partial v}{\partial y} \tag{4.61}$$

$$N_y = \lambda\left(\frac{\partial u}{\partial y} + \frac{\partial v}{\partial x}\right) \tag{4.62}$$

Substituting the assumed solutions given by Equations 4.52 and 4.53 into the above Equations followed by substitution of Equation 4.59, we get

$$\widehat{N}_x = C_1(-ik_1(\lambda + 2\mu)R_{11} - i\xi\lambda R_{21})e^{-ik_1 x}$$
$$+C_2(ik_1(\lambda + 2\mu)R_{12} - i\xi\lambda R_{22})e^{-ik_1(L-x)}$$
$$+C_3(-ik_2(\lambda + 2\mu)R_{13} - i\xi\lambda R_{23})e^{-ik_2 x}$$
$$+C_4(ik_2(\lambda + 2\mu)R_{14} - i\xi\lambda R_{24})e^{-ik_2(L-x)} \tag{4.63}$$
$$\widehat{N}_y = C_1\lambda(-i\xi R_{11} - ik_1 R_{21})e^{-ik_1 x} + C_2\lambda(-i\xi R_{12} + ik_1 R_{22})e^{-ik_1(L-x)}$$
$$C_3\lambda(-i\xi R_{13} - ik_2 R_{23})e^{-ik_2 x} + C_4\lambda(-i\xi R_{14} + ik_2 R_{24})e^{-ik_2(L-x)} \tag{4.64}$$

The above Equations can be written as a matrix equation of the form given by Equation 4.46 with the matrix $[\mathbf{R'}]$ having different form which can be obtained straightforwardly from Equations 4.63 and 4.64. Next, similar to the transformed nodal displacements, the transformed nodal forces denoted by the vector $\{\widehat{\mathbf{F}}^e\}$ can be related to the unknown constants $\{C\}$ as

$$\{\widehat{\mathbf{F}}^e\} = \begin{Bmatrix} \widehat{N}_{x1} \\ \widehat{N}_{y1} \\ \widehat{N}_{x2} \\ \widehat{N}_{y2} \end{Bmatrix} = [\mathbf{T_2}]\{C\} \tag{4.65}$$

The steps followed in deriving Equation 4.65 are similar to that adopted to derive Equation 4.48 and are not repeated here. Once Equations 4.60 and 4.65 are obtained, the dynamic stiffness matrix for the two noded FSFE of the two-dimensional waveguide can be calculated as in Equation 4.49.

Chapter 5

Wavelet Spectral Finite Element: Time Domain Analysis

This chapter discusses the development of a wavelet-based numerical technique for simulation of wave propagation in structural waveguides [62]. The method is based on wavelet transform and spectral finite element technique. The orthogonal, compactly supported Daubechies scaling functions are used for temporal approximation followed by finite element approach in the transformed domain. The method termed as *Wavelet Spectral Finite Element* (WSFE) is ideally suited for modeling finite dimension structures of relatively high complexities. This chapter is confined to the use of WSFE for modeling one-dimensional waveguide for time domain analysis. The following chapters describe the frequency domain analysis using WSFE and also extension to the modeling of two-dimensional structures.

The first example considered for illustrating the WSFE formulation is that of an isotropic rod. This is mainly to keep the mathematical expressions simple. Following this example, the case of an Euler-Bernoulli beam and higher order composite beam with elastic coupling [65, 68] is given.

5.1 Reduction of wave equations for a rod

The governing differential wave equation of an isotropic rod is given as [28]

$$EA\frac{\partial^2 u}{\partial x^2} - \eta A\frac{\partial u}{\partial t} = \rho A\frac{\partial^2 u}{\partial t^2} \qquad (5.1)$$

where E, A, η, and ρ are the Young's modulus, cross-sectional area, damping ratio, and density respectively. $u(x,t)$ is the axial deformation. Let $u(x,t)$ be discretized at n points in the time window $[0\ t_f]$. Let $\tau = 0,\ 1,\ldots,\ n-1$ be the sampling points, then

$$t = \Delta t \tau \qquad (5.2)$$

where Δt is the time interval between two sampling points. The function $u(x,t)$ can be approximated by scaling function $\varphi(\tau)$ at an arbitrary scale as

$$u(x,t) = u(x,\tau) = \sum_k u_k(x)\varphi(\tau - k), \qquad k \in \mathbf{Z} \qquad (5.3)$$

where $u_k(x)$ (referred as u_k hereafter) are the approximation coefficients at a certain spatial location x. Substituting Equations 5.2 and 5.3 in Equation 5.1 we get

$$EA\sum_k \frac{d^2 u_k}{dx^2}\varphi(\tau - k) - \frac{\eta A}{\Delta t}\sum_k u_k\varphi'(\tau - k) = \frac{\rho A}{\Delta t^2}\sum_k u_k\varphi''(\tau - k) \quad (5.4)$$

Taking inner product on both sides of Equation 5.4 with $\varphi(\tau - j)$, where $j = 0, 1, \ldots, n - 1$ we get

$$EA\sum_k \frac{d^2 u_k}{dx^2}\int \varphi(\tau - k)\varphi(\tau - j)d\tau - \frac{\eta A}{\Delta t}\sum_k u_k\int \varphi'(\tau - k)\varphi(\tau - j)d\tau$$
$$= \frac{\rho A}{\Delta t^2}\sum_k u_k\int \varphi''(\tau - k)\varphi(\tau - j)d\tau \qquad (5.5)$$

The translates of scaling functions are orthogonal, i.e.,

$$\int \varphi(\tau - k)\varphi(\tau - j)d\tau = 0 \quad \text{for} \quad j \neq k \qquad (5.6)$$

Using Equation 5.6, Equations 5.5 can be written as n simultaneous ODEs

$$EA\frac{d^2 u_j}{dx^2} - \frac{\eta A}{\Delta t}\sum_{k=j-N+2}^{j+N-2}\Omega^1_{j-k}u_k = \frac{\rho A}{\Delta t^2}\sum_{k=j-N+2}^{j+N-2}\Omega^2_{j-k}u_k$$
$$j = 0, 1, \ldots, n-1 \qquad (5.7)$$

$$EA\frac{d^2 u_j}{dx^2} = \sum_{k=j-N+2}^{j+N-2}\left(\frac{\eta A}{\Delta t}\Omega^1_{j-k} + \frac{\rho A}{\Delta t^2}\Omega^2_{j-k}\right)u_k$$
$$j = 0, 1, \ldots, n-1 \qquad (5.8)$$

where N, Ω^1_{j-k}, and Ω^2_{j-k} are the order of the Daubechies wavelet and connection coefficients as discussed earlier.

The forced boundary condition associated with the governing differential equation given by Equation 5.1 is

$$EA\frac{\partial u}{\partial x} = F \qquad (5.9)$$

where $F(x,t)$ is the axial force applied. $F(x,t)$ can be approximated similarly as $u(x,t)$ in Equation 5.3

$$F(x,t) = F(x,\tau) = \sum_k F_k(x)\varphi(\tau - k), \qquad k \in \mathbf{Z} \qquad (5.10)$$

Substituting Equations 5.3 and 5.10 in Equation 5.9 and taking the inner product with $\varphi(\tau - j)$ we get

$$EA\frac{du_j}{dx} = F_j \qquad j = 0,\ 1,\ldots, n-1 \tag{5.11}$$

As mentioned earlier in Chapter 3, for handling finite length data sequences, the coefficients u_j near the vicinity of the boundaries ($j = 0$ and $j = n - 1$) should be treated appropriately. Similar to the solution of ordinary differential equations for structural dynamics problems in Chapter 3, the wavelet extrapolation technique is used here to treat the boundary coefficients.

After treating the boundaries for analysis of the finite system, Equation 5.8 can be written in a matrix form,

$$\left\{\frac{d^2 u_j}{dx^2}\right\} = \left(\frac{\eta A}{EA}\Gamma^1 + \frac{\rho A}{EA}\Gamma^2\right)\{u_j\} \tag{5.12}$$

It should be noted that though all the formulations are done with reference to the governing differential equation for a rod, the connection coefficient matrices Γ^1 and Γ^2 are independent of the problem and depend only on the order of wavelet, i.e., N.

5.2   Decoupling using eigenvalue analysis

It can be seen from the above derivations that the wavelet coefficients of first and second derivatives can be obtained as

$$\{\dot{u}_j\} = \Gamma^1\{u_j\} \tag{5.13}$$

$$\{\ddot{u}_j\} = \Gamma^2\{u_j\} \tag{5.14}$$

The second derivative can also be written as

$$\{\ddot{u}_j\} = \Gamma^1\{\dot{u}_j\} \tag{5.15}$$

Substituting Equation 5.13 in Equation 5.15 we get

$$\{\ddot{u}_j\} = [\Gamma^1]^2\{u_j\} \tag{5.16}$$

Thus though the second order connection coefficient matrix Γ^2 can be evaluated independently [8], they can also be written as

$$\Gamma^2 = [\Gamma^1]^2 \tag{5.17}$$

The above modification is done as this form helps in imposing the initial conditions for non-periodic solution. Thus the Equation 5.12 can be written as

$$\left\{\frac{d^2 u_j}{dx^2}\right\} = \left(\frac{\eta A}{EA}[\Gamma^1] + \frac{\rho A}{EA}[\Gamma^1]^2\right)\{u_j\} \qquad (5.18)$$

In WSFE, the reduced ODEs are coupled; however, the system of equation can be decoupled by diagonalizing the connection coefficient matrix Γ^1. This can be done by eigenvalue analysis of the matrix as

$$\Gamma^1 = \Phi\Pi\Phi^{-1} \qquad (5.19)$$

where Φ is the eigenvector matrix of Γ^1 and Π is the diagonal matrix containing corresponding eigenvalues $-\iota\gamma_j$. From Equation 5.17, Γ^2 can be written as

$$\Gamma^2 = \Phi\Pi^2\Phi^{-1} \qquad (5.20)$$

where Π^2 is a diagonal matrix with diagonal terms $-\gamma_j^2$. This eigenvalue analysis is costly but can be done once and stored as it is completely independent of the problem. This makes the computational time comparable to Fourier transform-based methods.

The ODEs obtained by decoupling the Equation 5.18 can be written as

$$\frac{d^2 \widehat{u}_j}{dx^2} = -\left(\frac{\eta A}{EA}\iota\gamma_j + \frac{\rho A}{EA}\gamma_j^2\right)\widehat{u}_j \qquad j = 0,\ 1,\ldots,n-1 \qquad (5.21)$$

where

$$\widehat{u}_j = \Phi^{-1}u_j \qquad (5.22)$$

Similarly, the force boundary condition given by Equation 5.11 can be written as

$$\frac{d\widehat{u}_j}{dx} = \widehat{F}_j \qquad j = 0,\ 1,\ldots,n-1 \qquad (5.23)$$

where

$$EA\widehat{F}_j = \Phi^{-1}F_j \qquad (5.24)$$

5.3   Wavelet spectral finite element formulation for a rod

For the solution of the system of uncoupled ODEs given by Equation 5.21 a finite element approach is adopted in the transformed wavelet domain. These equations are required to be solved for $\widehat{u}_j$ and the actual solution $u(x,t)$ is obtained using inverse transform.

Figure 5.1(a) shows the spectral rod element with two nodes and one longitudinal degree of freedom (dof) $\widehat{u}_j$ and nodal axial load $\widehat{F}_j$ at each node.

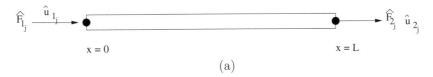

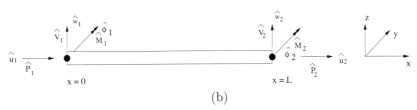

FIGURE 5.1: (a) Rod and (b) beam element with nodal forces and displacements

The formulation of 2×2 elemental dynamic stiffness matrix is explained as follows. Hereafter the subscript j is dropped for simplified notations and all the equations hold for $j = 0, 1, \ldots, n-1$. The exact interpolating function obtained by solving Equation 5.21 is

$$\widehat{u}(x) = C_1 e^{-\imath k x} + C_2 e^{-\imath k(L-x)} \quad \text{where} \quad k = \left(\frac{\eta A}{EA} \imath \gamma + \frac{\rho A}{EA} \gamma^2 \right)^{\frac{1}{2}} \quad (5.25)$$

Here, k is the wavenumber corresponding to the axial mode as explained in Chapter 3. The above equation can also be written in matrix form as

$$\widehat{u}(x) = [1 \quad 1] \begin{bmatrix} e^{-\imath k x} & 0 \\ 0 & e^{-\imath k(L-x)} \end{bmatrix} \begin{Bmatrix} C_1 \\ C_2 \end{Bmatrix}$$

$$\widehat{u}(x) = [R][\Theta]\{a\} \quad (5.26)$$

The constants $\{a\} = \{C_1 \ C_2\}^T$ are obtained from the boundary conditions at the two nodes as shown in Figure 5.1(a). The boundary conditions on the displacements can be written as

$$\widehat{u}(x)|_{@x=0} = \widehat{u}_1 = [R] \begin{bmatrix} 1 & 0 \\ 0 & e^{-\imath k L} \end{bmatrix} \begin{Bmatrix} C_1 \\ C_2 \end{Bmatrix}$$

$$= [T_{11}]\{a\} \quad (5.27)$$

$$\widehat{u}(x)|_{@x=L} = \widehat{u}_2 = [R] \begin{bmatrix} e^{-\imath k L} & 0 \\ 0 & 1 \end{bmatrix} \begin{Bmatrix} C_1 \\ C_2 \end{Bmatrix}$$

$$= [T_{12}]\{a\} \quad (5.28)$$

Combining Equations 5.27 and 5.28, the nodal displacement vector can be written as

$$\{\widehat{u}^e\} = \begin{Bmatrix} \widehat{u}_1 \\ \widehat{u}_2 \end{Bmatrix} \begin{bmatrix} T_{11} \\ T_{12} \end{bmatrix} \{a\} = [T_1]\{a\} \tag{5.29}$$

Similarly, from Equation 5.24, the force boundary condition can be written as

$$\widehat{F}(x) = EA(-\imath k C_1 e^{-\imath k x} + \imath k C_2 e^{-\imath k(L-x)}) \tag{5.30}$$

Writing the above equation in a matrix form,

$$\widehat{F}(x) = EA[-\imath k \quad \imath k] \begin{bmatrix} e^{-\imath k x} & 0 \\ 0 & e^{-\imath k(L-x)} \end{bmatrix} \begin{Bmatrix} C_1 \\ C_2 \end{Bmatrix}$$

$$= [\acute{R}][\Theta]\{a\} \tag{5.31}$$

Next, applying the boundary conditions at the two nodes,

$$\widehat{F}(x)|_{@x=0} = \widehat{F}_1 = [\acute{R}] \begin{bmatrix} 1 & 0 \\ 0 & e^{-\imath k L} \end{bmatrix} \begin{Bmatrix} C_1 \\ C_2 \end{Bmatrix}$$

$$= [T_{21}]\{a\} \tag{5.32}$$

$$\widehat{F}(x)|_{@x=L} = \widehat{F}_2 = [\acute{R}] \begin{bmatrix} e^{-\imath k L} & 0 \\ 0 & 1 \end{bmatrix} \begin{Bmatrix} C_1 \\ C_2 \end{Bmatrix}$$

$$= [T_{22}]\{a\} \tag{5.33}$$

Combining Equations 5.32 and 5.33, the nodal force vector can be written as

$$\{\widehat{F}^e\} = \begin{Bmatrix} \widehat{F}_1 \\ \widehat{F}_2 \end{Bmatrix} = \begin{bmatrix} T_{21} \\ T_{22} \end{bmatrix} \{a\} = [T_2]\{a\} \tag{5.34}$$

Eliminating $\{a\}$ from Equations 5.29 and 5.34, the nodal displacement vector can be related to the nodal force vector as

$$\{\widehat{F}^e\} = [T_2][T_1]^{-1}\{\widehat{u}^e\} = [\widehat{K}^e]\{\widehat{u}^e\} \tag{5.35}$$

Here, $[\widehat{K}^e]$ is the elemental dynamic stiffness matrix. The calculations of the above matrices for $j = 0, 1, \ldots, n-1$ are done numerically. The above equation can be solved to obtain the nodal displacements $\{\widehat{u}^e\}$ from the given nodal forces $\{\widehat{u}^e\}$ after imposing the boundary conditions similar to FE. After $\{\widehat{u}^e\}$, the constants $\{a\}$ can be derived from Equation 5.29 as

$$\{a\} = [T_1]^{-1}\{\widehat{u}^e\} \tag{5.36}$$

Substituting $\{a\}$ into Equation 5.26 will allow to obtain the response $\widehat{u}(x)$ at any arbitrary point within the rod. The calculation of stresses and strains is obtained using standard finite element procedure.

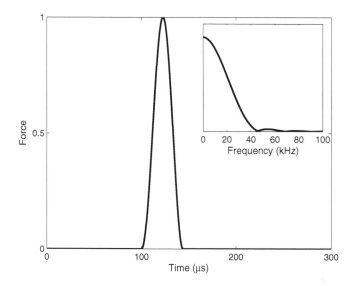

FIGURE 5.2:   Impact load and Fourier transform of the load (inset)

5.4   Time domain response of elementary rod under impulse load

Here, first the formulated wavelet spectral element is compared with 2-D FE. Next, numerical experiments are presented to highlight the advantages of WSFE over the corresponding Fourier transform-based method, particularly for modeling finite length structures.

5.4.1   Comparison with finite element solution

First, wave propagation analysis is done for an aluminum rod with Young's modulus $E = 70$ GPa and density $\rho = 2.7 \times 10^3$ kg/m^3. The rod is fixed at one end and an axial impulse load as shown in Figure 5.2 is applied at the free end. The load has an unit amplitude and duration of 50 μs with a frequency content of 44 kHz. The length, width, and depth of the rod are $L = 20$ in, $b = 1$ in and $h = 0.01$ in respectively.

Figure 5.3(a) shows the tip longitudinal velocity in undamped ($\eta = 0$) rod due to impact load applied at tip. The result obtained using a single formulated spectral element is compared with 2-D FE result. The FE result is obtained using 400 3-noded plane stress triangular elements and Newmark's time integration. The wavelet basis function used in this example has an order of $N = 22$ and the sampling rate $\Delta t = 1$ μs. Thus for a time window

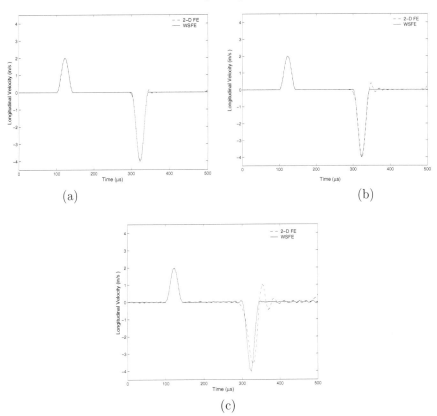

FIGURE 5.3:   Longitudinal tip velocity in rod due to tip impact load simulated with time interval (a) $\triangle t = 1$ μs, (b) $\triangle t = 2$ μs, and (c) $\triangle t = 4$ μs

$T_w = 512$ μs, the number of sampling points is $n = 512$ and the system size is 512×512. It can be seen that the FE results match well with those obtained using formulated WSFE.

In Figures 5.3(b) and (c), the FE results obtained using the above mesh (400 3-noded plane stress triangular elements) but with time steps of 2 μs and 4 μs for Newmark's time integration are presented. These are compared with corresponding WSFE results obtained with $\triangle t = 2$ μs and 4 μs respectively. The time window T_w is kept fixed at 512 μs. It can be seen from Figures 5.3(a) to (c) that as the time step is increased, the FE method with Newmark's time integration scheme gradually shows distortions. However, WSFE is free from such distortions even at $\triangle t = 4$ μs.

In Table 5.1, the CPU time taken by FE and WSFE methods for the above numerical examples is presented for $\triangle t = 1$ μs, 2 μs, and 4 μs. The numerical experiments are performed on an IBM IntelliStation workstation. It can be seen that the CPU time taken by the present WSFE method is much less than that taken by FE method. In addition, it can be seen from Fig 5.3(c) that

TABLE 5.1: CPU time taken by FE with Newmark's time integration and WSFE methods for simulation of longitudinal tip velocity in rod shown in Figure 5.3

Time Interval $\triangle t$ (in μs)	CPU time (in sec)	
	2-D FE	Present (WSFE)
1	54.00	18.00
2	27.00	2.00
4	15.00	1.00

for $\triangle t = 4$ μs, WSFE solution is much more accurate than the corresponding FE solution. This leads to further computational savings in WSFE compared over FE analysis with Newmark's time integration.

5.4.2 Advantage over Fourier transform-based spectral finite element

Wave propagation analysis of finite length structures using conventional SFE based on Fourier transform requires the structures to be damped or the use of throw off element [36] to artificially induce damping. In addition, for such methods, the time window should be large to remove the wrap around problem. The time window is dependent on the value of damping and the length of the structure, and requires to be more for lightly damped short length structures. WSFE is completely free from such constraints where the accuracy of solution is independent of these parameters. In Figure 5.4 longitudinal velocities in the rod due to tip impact load are plotted considering $\eta = 0.5$. For the solution obtained using WSFE, the time window of $T_w = 512$ μs is sufficient, while for FSFE time windows, T_w of 1024 μs, 2048 μs and 4096 μs and $\Delta t = 1$ μs are used. It can be seen that for $T_w = 1024$ μs, FSFE solution is highly distorted and the accuracy gradually increases with increase of T_w (Figure 5.4(c)). Thus, the present spectral element results in substantial reduction of computational cost as the T_w is directly related to the system size.

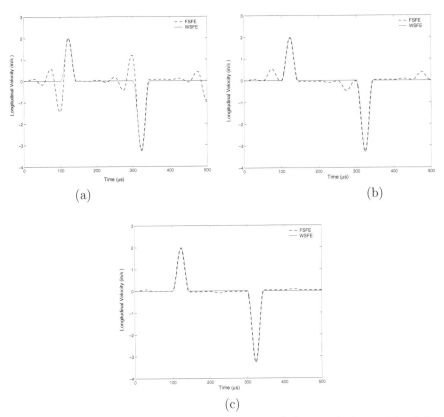

FIGURE 5.4:   Longitudinal tip velocity in rod due to tip impact load for time window T_w (a) $T_w = 1024\ \mu s$, (b) $T_w = 2048\ \mu s$, and (c) $T_w = 4096\ \mu s$

5.5   Reduction of wave equations for Euler-Bernoulli beam

The flexural wave equation for an Euler-Bernoulli beam has a fourth order derivative in space and is given as [28]

$$EI\frac{\partial^4 w}{\partial x^4} + \eta A\frac{\partial w}{\partial t} + \rho A\frac{\partial^2 w}{\partial t^2} = 0 \qquad (5.37)$$

where $w(x,t)$ is the transverse displacement and I is the moment of inertia of the cross-section. The main difference between longitudinal and flexural waves is that the latter is dispersive; in other words the wave speeds vary with frequencies.

Similar to the approximation of the longitudinal displacement $u(x,t)$ given by Equation 5.3, the transverse displacement $w(x,t)$ is approximated as

$$w(x,t) = w(x,\tau) = \sum_k w_k(x)\varphi(\tau - k), \qquad k \in \mathbf{Z} \qquad (5.38)$$

where w_k are the approximation coefficients. Substituting Equation 5.38 in Equation 5.37 and following similar steps of Equations 5.4 to 5.6, we get the reduced ODEs as

$$EI\frac{d^4 w_j}{dx^4} + \frac{\eta A}{\Delta t}\sum_{k=j-N+2}^{j+N-2}\Omega^1_{j-k}w_k + \frac{\rho A}{\Delta t^2}\sum_{k=j-N+2}^{j+N-2}\Omega^2_{j-k}w_k = 0$$
$$j = 0, 1, \ldots, n-1 \qquad (5.39)$$

$$EI\frac{d^4 w_j}{dx^4} + \sum_{k=j-N+2}^{j+N-2}\left(\frac{\eta A}{\Delta t}\Omega^1_{j-k} + \frac{\rho A}{\Delta t^2}\Omega^2_{j-k}\right)w_k = 0$$
$$j = 0, 1, \ldots, n-1 \qquad (5.40)$$

The forced boundary conditions associated with the governing equation (Equation 5.37) are

$$EI\frac{\partial^2 w}{\partial x^2} = M \qquad (5.41)$$

$$EI\frac{\partial^3 w}{\partial x^3} = -V \qquad (5.42)$$

where M and V are the applied moment and transverse force respectively. Similar to Equation 5.10, $M(x,t)$ and $V(x,t)$ are written as

$$M(x,t) = M(x,\tau) = \sum_k M_k(x)\varphi(\tau - k), \qquad k \in \mathbf{Z} \qquad (5.43)$$

$$V(x,t) = V(x,\tau) = \sum_k V_k(x)\varphi(\tau - k), \qquad k \in \mathbf{Z} \qquad (5.44)$$

Substituting Equations 5.43 and 5.44 in Equations 5.41 and 5.42 respectively we get the following ODEs

$$EI\frac{d^2 w_j}{dx^2} = M_j \qquad j = 0,\ 1,\ldots,n-1 \qquad (5.45)$$

$$EI\frac{d^3 w_j}{dx^3} = -V_j \qquad j = 0,\ 1,\ldots,n-1 \qquad (5.46)$$

Spectral element for beam is formulated using the ODEs given by Equations 5.40, 5.45, and 5.46.

Next, the coefficients lying outside the finite boundaries are treated through wavelet extrapolation technique. Thus the ODEs given by Equation 5.40 can be written as a matrix equation of the form

$$\left\{\frac{d^4 w_j}{dx^4}\right\} + \left(\frac{\eta A}{EI}\Gamma^1 + \frac{\rho A}{EI}\Gamma^2\right)\{w_j\} = 0 \qquad (5.47)$$

Again, here, the second order connection coefficient matrix Γ^2 is replaced by $[\Gamma^1]^2$, as given by Equation 5.17. This modification results in the following equation from Equation 5.47

$$\left\{\frac{d^4 w_j}{dx^4}\right\} + \left(\frac{\eta A}{EI}[\Gamma^1] + \frac{\rho A}{EI}[\Gamma^1]^2\right)\{w_j\} = 0 \qquad (5.48)$$

The decoupling of the Equation 5.47 is done through eigenvalue analysis and the decoupled equation is obtained as

$$\frac{d^4 \widehat{w}_j}{dx^4} + \left(\frac{\eta A}{EI}\gamma_j + \frac{\rho A}{EI}\gamma_j^2\right)\widehat{w}_j = 0 \qquad j = 0,\ 1,\ldots,n-1 \qquad (5.49)$$

where

$$\widehat{w}_j = \Phi^{-1} w_j \qquad (5.50)$$

Φ is the eigenvector matrix and $\imath\gamma_j$ are the eigenvalues of the matrix Γ^1 as given by Equations 5.19 and 5.20.

The forced boundary conditions given in Equations 5.45 and 5.46 are similarly transformed as

$$EI\frac{d^2 \widehat{w}_j}{dx^2} = \widehat{M}_j \qquad j = 0,\ 1,\ldots,n-1 \qquad (5.51)$$

$$EI\frac{d^3 \widehat{w}_j}{dx^3} = -\widehat{V}_j \qquad j = 0,\ 1,\ldots,n-1 \qquad (5.52)$$

5.6   WSFE formulation for Euler-Bernoulli beam

Figure 5.1(b) shows the beam spectral finite element with two dofs, $\widehat{w}$ and $\partial\widehat{w}/\partial x$ represented as θ in Figure 5.1(b), at each node. The nodal transverse forces and moments are $\widehat{V}$ and $\widehat{M}$ respectively. Spectral element formulation for beam is very similar to that for rod. The solution of Equation 5.49 is

$$\widehat{w}(x) = C_1 e^{-\imath kx} + C_2 e^{-\imath k(L-x)} + C_3 e^{-kx} + C_4 e^{-k(L-x)}$$

$$\text{where } k = \left(\frac{\eta A}{EA}\imath\gamma + \frac{\rho A}{EA}\gamma^2\right)^{\frac{1}{4}} \tag{5.53}$$

Here, k are the wavenumbers and $\{a\} = \{C_1\ C_2\ C_3\ C_4\}^T$ are the constants to be derived from the boundary conditions at the two nodes. Writing the above equation in matrix form, we get

$$\widehat{w}(x) = [1\ 1\ 1\ 1] \begin{bmatrix} e^{-\imath kx} & 0 & 0 & 0 \\ 0 & e^{-\imath k(L-x)} & 0 & 0 \\ 0 & 0 & e^{-kx} & 0 \\ 0 & 0 & 0 & e^{-k(L-x)} \end{bmatrix} \begin{Bmatrix} C_1 \\ C_2 \\ C_3 \\ C_4 \end{Bmatrix}$$

$$= [R_1][\Theta]\{a\} \tag{5.54}$$

Correspondingly, the rotational dof $\partial \widehat{w}/\partial x$ can be written as

$$\frac{\partial \widehat{w}(x)}{\partial x} = [-\imath k\ \imath k\ -k\ k] \begin{bmatrix} e^{-\imath kx} & 0 & 0 & 0 \\ 0 & e^{-\imath k(L-x)} & 0 & 0 \\ 0 & 0 & e^{-kx} & 0 \\ 0 & 0 & 0 & e^{-k(L-x)} \end{bmatrix} \begin{Bmatrix} C_1 \\ C_2 \\ C_3 \\ C_4 \end{Bmatrix}$$

$$= [R_2][\Theta]\{a\} \tag{5.55}$$

Applying the boundary conditions at the two nodes, we get

$$\widehat{w}(x)|_{@x=0} = \widehat{w}_1 = [R_1] \begin{bmatrix} 1 & 0 & 0 & 0 \\ 0 & e^{-\imath kL} & 0 & 0 \\ 0 & 0 & 1 & 0 \\ 0 & 0 & 0 & e^{-kL} \end{bmatrix} \begin{Bmatrix} C_1 \\ C_2 \\ C_3 \\ C_4 \end{Bmatrix}$$

$$= [T_{11}]\{a\} \tag{5.56}$$

$$\frac{\partial \widehat{w}(x)}{\partial x}\bigg|_{@x=0} = \widehat{\theta}_1 = [R_2] \begin{bmatrix} 1 & 0 & 0 & 0 \\ 0 & e^{-\imath kL} & 0 & 0 \\ 0 & 0 & 1 & 0 \\ 0 & 0 & 0 & e^{-kL} \end{bmatrix} \begin{Bmatrix} C_1 \\ C_2 \\ C_3 \\ C_4 \end{Bmatrix}$$

$$= [T_{12}]\{a\} \tag{5.57}$$

$$\widehat{w}(x)|_{@x=L} = \widehat{w}_2 = [R_1] \begin{bmatrix} e^{-\imath kL} & 0 & 0 & 0 \\ 0 & 1 & 0 & 0 \\ 0 & 0 & e^{-kL} & 0 \\ 0 & 0 & 0 & 1 \end{bmatrix} \begin{Bmatrix} C_1 \\ C_2 \\ C_3 \\ C_4 \end{Bmatrix}$$

$$= [T_{13}]\{a\} \tag{5.58}$$

$$\frac{\partial \widehat{w}(x)}{\partial x}\Big|_{@x=L} = \widehat{\theta}_2 = [R_2] \begin{bmatrix} e^{-\imath kL} & 0 & 0 & 0 \\ 0 & 1 & 0 & 0 \\ 0 & 0 & e^{-kL} & 0 \\ 0 & 0 & 0 & 1 \end{bmatrix} \begin{Bmatrix} C_1 \\ C_2 \\ C_3 \\ C_4 \end{Bmatrix}$$

$$= [T_{14}]\{a\} \tag{5.59}$$

Combining Equations 5.56 to 5.59, the nodal displacement vector can be written as

$$\{\widehat{u}^e\} = \begin{Bmatrix} \widehat{w}_1 \\ \widehat{\theta}_1 \\ \widehat{w}_2 \\ \widehat{\theta}_2 \end{Bmatrix} = \begin{bmatrix} \mathbf{T}_{11} \\ \mathbf{T}_{12} \\ \mathbf{T}_{13} \\ \mathbf{T}_{14} \end{bmatrix} \{a\}$$

$$\{\widehat{u}^e\} = [T_1]\{a\} \tag{5.60}$$

Similar boundary conditions can be applied to the shear force $\widehat{V}(x)$ and moment $\widehat{M}(x)$, given by Equations 5.51 and 5.52, which can written in matrix forms as

$$\widehat{V}(x) = -EI(\imath k^3 C_1 e^{-\imath kx} - \imath k^3 C_2 e^{-\imath k(L-x)} - k^3 C_3 e^{-kx} + k^3 C_4 e^{-k(L-x)})$$

$$\widehat{V}(x) = -EI[\imath k^3 \ -\imath k^3 \ -k^3 \ k^3] \begin{bmatrix} e^{-\imath kx} & 0 & 0 & 0 \\ 0 & e^{-\imath k(L-x)} & 0 & 0 \\ 0 & 0 & e^{-kx} & 0 \\ 0 & 0 & 0 & e^{-k(L-x)} \end{bmatrix} \begin{Bmatrix} C_1 \\ C_2 \\ C_3 \\ C_4 \end{Bmatrix}$$

$$= [\acute{R}_1][\Theta]\{a\} \tag{5.61}$$

$$\widehat{M}(x) = EI(-k^2 C_1 e^{-\imath kx} + k^2 C_2 e^{-\imath k(L-x)} + k^2 C_3 e^{-kx} + k^2 C_4 e^{-k(L-x)})$$

$$\widehat{M}(x) = EI[k^2 \ k^2 \ k^2 \ k^2] \begin{bmatrix} e^{-\imath kx} & 0 & 0 & 0 \\ 0 & e^{-\imath k(L-x)} & 0 & 0 \\ 0 & 0 & e^{-kx} & 0 \\ 0 & 0 & 0 & e^{-k(L-x)} \end{bmatrix} \begin{Bmatrix} C_1 \\ C_2 \\ C_3 \\ C_4 \end{Bmatrix}$$

$$= [\acute{R}_2][\Theta]\{a\} \tag{5.62}$$

The boundary values of the above force and moment at the two nodes of the spectral element can be written as

$$\widehat{V}(x)|_{@x=L} = \widehat{V}_2 = [\acute{R}_1] \begin{bmatrix} e^{-\imath kL} & 0 & 0 & 0 \\ 0 & 1 & 0 & 0 \\ 0 & 0 & e^{-kL} & 0 \\ 0 & 0 & 0 & 1 \end{bmatrix} \{a\}$$

$$\widehat{V}_2 = [T_{23}]\{a\} \tag{5.63}$$

$$\widehat{M}(x)|_{@x=L} = \widehat{M}_2 = [\acute{R}_2] \begin{bmatrix} e^{-\imath kL} & 0 & 0 & 0 \\ 0 & 1 & 0 & 0 \\ 0 & 0 & e^{-kL} & 0 \\ 0 & 0 & 0 & 1 \end{bmatrix} \{a\}$$

$$\widehat{V}_1 = [T_{24}]\{a\} \tag{5.64}$$

Augmenting Equations 5.61 to 5.64, the nodal force vector can be written as

$$\{\widehat{F}^e\} = \left\{ \begin{array}{c} \widehat{V}_1 \\ \widehat{M}_1 \\ \widehat{V}_2 \\ \widehat{M}_2 \end{array} \right\} = \begin{bmatrix} \mathbf{T}_{21} \\ \mathbf{T}_{22} \\ \mathbf{T}_{23} \\ \mathbf{T}_{24} \end{bmatrix} \{a\}$$

$$\{\widehat{F}^e\} = [T_2]\{a\} \tag{5.65}$$

Finally, eliminating $\{a\}$ from Equations 5.60 and 5.65, the nodal displacements can be related to the nodal forces as

$$\{\widehat{F}^e\} = [T_2][T_1]^{-1}\{\widehat{u}^e\} = [\widehat{K}^e]\{\widehat{u}^e\} \tag{5.66}$$

Here, $[\widehat{K}^e]$ is the elemental dynamic stiffness matrix for Euler-Bernoulli beam. The above equation can be solved to obtain the nodal displacement vector, $\{\widehat{u}^e\}$ for known nodal forces. $\{\widehat{u}^e\}$ can be substituted into Equation 5.60 to derive the constants $\{a\}$ as

$$\{a\} = [T_1]^{-1}\{\widehat{u}^e\} \tag{5.67}$$

Knowing $\{a\}$, it can be substituted to Equations 5.54 and 5.55 to obtain the displacements $\widehat{w}(x)$ and $\frac{\partial \widehat{w}(x)}{\partial x}$ at any arbitrary point on the beam.

Matlab program 5.1 gives the Matlab code to develop the WSFE for isotropic Euler-Bernoulli beam along with axial displacement. The code uses the function *wsfe_ebt* given as Matlab program 5.2. Therefore, the two noded element has three dofs at each node, that are axial, transverse, and rotational. Though in the above formulation shown for elementary rod and Euler-Bernoulli beam, the wavenumbers k are obtained analytically, in the code given, they are obtained numerically using the Matlab function *root*. The matrices $[R_1]$ and $[R_2]$ are also obtained through SVD as explained in Chapter 3. The code uses several functions other than Matlab in-built functions. These functions have been presented in Chapter 3.

5.7   Time domain response of Euler-Bernoulli beam under impulse load

The flexural wave in isotropic beam modeled using Euler-Bernoulli theory is simulated and validated with 2-D FE results. Comparison is also done with the corresponding FSFE simulation results.

Numerical experiments are performed to study flexural wave propagation in an aluminum beam due to the unit impulse load applied at tip in transverse

```
-----------------------------------------------------
MATLAB Program 5.1
-----------------------------------------------------
% INPUT SIGNAL
fname=input('Enter input file name: ','s');
pulse=load(fname);
t=pulse(:,1); F=pulse(:,2);
t_s=t(2)-t(1); % sampling time
m=10; % level of resolution
n=2^m; % No. of wavelet coefficient
if n>length(F)
F(s+1:n)=0;% zero padding of the load vector
else F=F(1:n); end
% MATERIAL PROPERTIES & DIMENSIONS of the aluminum beam
E=70e+09; rho=2700; % Young's modulus & mass density
b=0.05; h=0.01;   % width & depth of the beam
L=1.0; xs=1.0     % length of beam & point of measurement
A=b*h; I=(1/12)*b*h^3;   % c/s area & moment of inertia
I_0=rho*A;        % Inertial constant
A_11=E*A; D_11=E*I; % Axial & Flexural stiffness constant
% DAUBECHIES BASIS function
DL=22;  d=1;  % orders of the basis function & derivative
% CONNECTION COEFF. through Wavelet Extrapolation technique
CC1=wavenp(DL,n,d);  % Function "wavenp" given in
% Chapter 3 [CC1]=[Gamma_1]
[V,D]=eig(CC1);       % Eigenvalue analysis of the matrix
DS=j*(1/dt)*diag(D); % Eigenvalues gamma_j
FN=inv(V)*F; % Transformation of the force
for i=1:n
    lambda=DS(i);
    % ELEMENTAL DYNAMIC STIFFNESS MATRIX KD
    [KD,SA,SB]=wsfe_ebt(I_0,A_11,D_11,L,xs,lambda);
    % Imposition of boundary condition for FIXED-FREE beam
    KD(1:3,:)=[]; KD(:,1:3)=[];
    % Nodal displacements due to axial loading
    ua_nodal=[FN(i,1) 0 0]*inv(KD);
    UA=[zeros(1,3) ua_nodal].';
    % Nodal displacements due to transverse loading
    ub_nodal=[0 FN(i,1) 0]*inv(KD);
    UB=[zeros(1,3) ub_nodal].';
    va_t(i,1)=SA*UA;% Axial Velocity
    vb_t(i,1)=SB*UB;% Transverse velocity
end
% INVERSE TRANSFORM to obtain time domain response
va=V*va_t; % Axial velocity in time domain
vb=V*vb_t; % Transverse velocity in time domain
```

```
-------------------------------------------------------
MATLAB Program 5.2
-------------------------------------------------------
function [KD,SA,SB]=wsfe_ebt(I0,A11,D11,L,xs,lambda)
CONST_R=zeros(2,6); CONST_N=zeros(2,6); CONST_M=zeros(3,6);
C(1)=A11*D11; C(3)=-lambda^2*I0*D11; C(5)=-lambda^2*I0*A11;
C(7)=lambda^4*I0^2; kj=roots(C);
    for p=1:6
        X=[(-lambda^2*I0+kj(p)^2*A11) 0; 0
        (-lambda^2*I0+kj(p)^4*D11)]; [UU,S,VV] = svd(X);
        CONST_R(:,p)=VV(:,2);
    end
[Y,I]=sort(abs(real(kj)));
    for l=1:6
        CONST_N(:,l)=CONST_R(:,I(l)); k_j(l)=kj(I(l));
    end
ih=1;
for pp=1:3
    if sign(imag(k_j(ih)))==1
        TEMP=CONST_N(:,ih); CONST_N(:,ih)=CONST_N(:,ih+1);
        CONST_N(:,ih+1)=TEMP;
        temp=k_j(ih); k_j(ih)=k_j(ih+1); k_j(ih+1)=temp;
    end
    ih=ih+2;
end
EL1=exp(-j*k_j(1)*L);EL2=exp(-j*k_j(3)*L);EL3=exp(-j*k_j(5)*L);
CONST_M(1:2,:)=CONST_N(1:2,:);
    for ic=1:6
        CONST_M(3,ic)=-j*k_j(ic)*CONST_N(2,ic);
    end
DG1=diag([1;EL1;1;EL2;1;EL3]); T11=CONST_M*DG1;
DG2=diag([EL1;1;EL2;1;EL3;1]); T12=CONST_M*DG2; T1=[T11;T12];
foril=1:6
P(1,il)=A11*(-j*k_j(il)*CONST_N(1,il));   % Axial load P
T(1,il)=D11*(-j*k_j(il)^3*CONST_N(2,il)); % Bending Load T
M(1,il)=-k_j(il)^2*D11*CONST_N(2,il);     % Bending Moment M
end
P1=P*DG1; P2=P*DG2; V1=T*DG1; V2=T*DG2; M1=M*DG1; M2=M*DG2;
T2=[P1;V1;M1;-P2;-V2;-M2];
KD=T2*inv(T1); % elemental dynamic stiffness matrix
DXS=diag([exp(-j*k_j(1)*xs) exp(-j*k_j(1)*(L-xs))...
exp(-j*k_j(3)*xs) exp(-j*k_j(3)*(L-xs))...
exp(-j*k_j(5)*xs)exp(-j*k_j(5)*(L-xs))]);
SA=(j*lambda)*CONST_M(1,:)*DXS*inv(T1);
SB=(j*lambda)*CONST_M(2,:)*DXS*inv(T1);
return
```

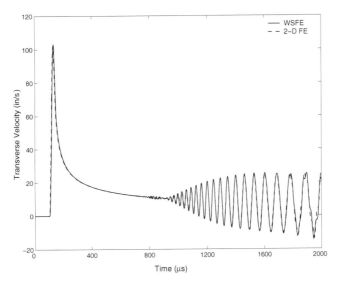

FIGURE 5.5:   Transverse tip velocity in beam due to tip impact load

direction. The elastic properties and dimensions are same as the rod except that here the length L is even shorter and is equal to 10 in. The transverse velocity at the tip simulated using a single WSFE is shown in Figure 5.5. A time window of $T_w = 2048$ μs with a sampling rate $\triangle t = 1$ μs is used to solve the problem. It is also compared with FE result obtained using 400, 3 noded plane stress triangular elements. Figure shows excellent agreement between FE and WSFE solutions.

In Figure 5.6, transverse velocity in the beam due to tip impact load is plotted considering a damping of $\eta = 0.5$. For the results obtained with WSFE, the time window used is $T_w = 1024$ μs; however as stated earlier the accuracy of the solution is independent of T_w. For conventional FSFE solutions, the time windows, T_w, used are 1024 μs, 2048 μs, and 4096 μs. It can be seen from Figures 5.6 (a) to (c) that the distortions gradually decrease by increasing T_w.

5.8   Wave propagation in frame structure

The elemental dynamic stiffness matrix $[\widehat{\mathbf{K}}]$ derived for the isotropic rod and beam in Sections 5.3 and 5.6 relates the transformed nodal displacements to the transformed nodal forces. In spectral finite element formulation, one element is sufficient to accurately model a structure of arbitrary length in absence of discontinuity. In order to model further complex structures with

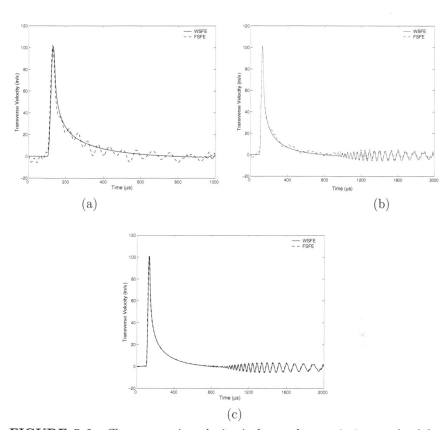

FIGURE 5.6: Transverse tip velocity in beam due to tip impact load for time window T_w (a) $T_w = 1024\ \mu s$, (b) $T_w = 2048\ \mu s$, and (c) $T_w = 4096\ \mu s$

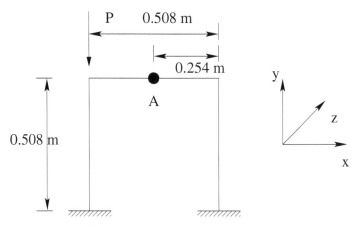

FIGURE 5.7: 2-D frame structure

joints and other discontinuities like that of stepped rod, bi-material structure, 2-D frame structure, the elemental structure has to be assembled to obtain the global dynamic stiffness matrix of such build-up structure.

The example of a 2-D frame structure with a vertical P shown in Figure 5.7 is considered to show the use of WSFE for simulation of wave propagation in relatively complex structures. This example is much more complicated as multiple reflections occur from the joints and supports. For modeling of this structure, three spectral finite elements are required. Each element should have three dofs, namely, axial $\hat{u}$, transverse $\hat{w}$, and rotational $\partial\hat{w}/\partial x$. Such an element can be obtained by combining the elements obtained separately for rod and Euler-Bernoulli beam in Sections 5.3 and 5.6. In Figure 5.8 the transverse wave velocity at point A is presented and compared with FE result. The FE result is obtained using 2-noded one-dimensional beam element with axial, transverse, and rotational degrees of freedom at each node. Each of the three members of the frame is discretized with 5000 elements. Results show excellent agreement between these two solutions. Since FSFE cannot be used for similar analysis of undamped finite length structures, the wave velocities are plotted in Figures 5.9(a) to (c) considering a damping of $\eta = 0.5$. In the above mentioned figures for all the plots obtained using wavelet, a time window T_w of 1024 μs is used. For FSFE, T_w is increased from 1024 μs to 4096 μs in Figures 5.9(a) to (c) to remove distortions due to wrap around. It can be seen that for $T_w = 1024$ μs the results are highly distorted which gradually decreases with increase in T_w. In Figure 5.9(c), even increasing T_w to 4096 μs is not capable of completely eliminating the response distortion and requires higher resolution.

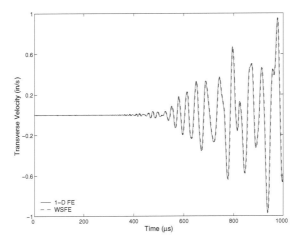

FIGURE 5.8:   Transverse velocity at A of 2-D frame in Figure 5.7 due to the applied load P

5.9   Governing differential wave equations for higher order composite beam

The governing differential wave equations for higher order composite beam are [56]

$$I_0\frac{\partial^2 u}{\partial t^2} - A_{11}\frac{\partial^2 u}{\partial x^2} + B_{11}\frac{\partial^2 \phi}{\partial x^2} - A_{13}\frac{\partial \psi}{\partial x} = 0 \tag{5.68}$$

$$I_2\frac{\partial^2 \psi}{\partial t^2} + A_{13}\frac{\partial u}{\partial x} - B_{13}\frac{\partial \phi}{\partial x} + A_{33}\psi$$

$$-B_{55}\left(\frac{\partial^2 w}{\partial x^2} - \frac{\partial \phi}{\partial x}\right) - D_{55}\frac{\partial^2 \psi}{\partial x^2} = 0 \tag{5.69}$$

$$I_0\frac{\partial^2 w}{\partial t^2} - A_{55}\left(\frac{\partial^2 w}{\partial x^2} - \frac{\partial \phi}{\partial x}\right) - B_{55}\frac{\partial^2 \psi}{\partial x^2} = 0 \tag{5.70}$$

$$I_2\frac{\partial^2 \phi}{\partial t^2} - A_{55}\left(\frac{\partial w}{\partial x} - \phi\right) - B_{55}\frac{\partial \psi}{\partial x} + B_{11}\frac{\partial^2 u}{\partial x^2}$$

$$-D_{11}\frac{\partial^2 \phi}{\partial x^2} + B_{11}\frac{\partial \psi}{\partial x} = 0 \tag{5.71}$$

Here, $\phi(x,t)$ and $\psi(x,t)$ are the additional dofs for higher order beam and represent the shear and contractional displacements as shown in Figure 5.10(a). The stiffness coefficients are functions of ply properties, orientations, etc. and

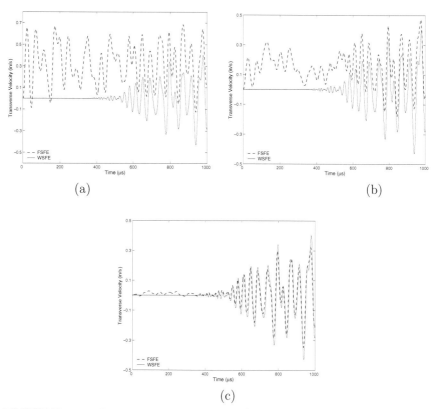

(a)

(b)

(c)

FIGURE 5.9:   Transverse tip velocity at A of 2-D frame in Figure 5.7 due to the applied load P, for time windows T_w (a) $T_w = 1024\ \mu s$, (b) $T_w = 2048\ \mu s$, and (c) $T_w = 4096\ \mu s$

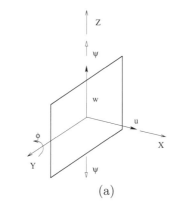

(a)

(b)

FIGURE 5.10: (a) Beam cross-section and the displacements; (b) composite beam element with nodal displacements and forces

are obtained by integrating over the beam cross-section as

$$[A_{ij}, B_{ij}, D_{ij}] = \sum \int_{z_i}^{z_{i+1}} \bar{Q}_{ij}[1, z, z^2]bdz \qquad (5.72)$$

where $\bar{Q}_{ij}$ are the transformed stiffness coefficients of each ply. z_i and z_{i+1} are the thickness coordinates of the ith layer and b is the corresponding width. Similarly, the inertial constants are obtained as

$$[I_0, I_2] = \sum \int_{z_i}^{z_{i+1}} \rho[1, z^2]bdz \qquad (5.73)$$

where ρ is the mass density. The force boundary conditions associated with the governing Equations 5.68 to 5.71 are

$$A_{11}\frac{\partial u}{\partial x} - B_{11}\frac{\partial \phi}{\partial x} + A_{13}\psi = P \qquad (5.74)$$

$$A_{55}\frac{\partial w}{\partial x} + B_{55}\frac{\partial \psi}{\partial x} - A_{55}\phi = V \qquad (5.75)$$

$$-B_{11}\frac{\partial u}{\partial x} + D_{11}\frac{\partial \phi}{\partial x} - B_{13}\psi = M \qquad (5.76)$$

$$B_{55}\frac{\partial w}{\partial x} + D_{55}\frac{\partial \psi}{\partial x} - B_{55}\phi = Q \qquad (5.77)$$

5.10 WSFE formulation for composite beam

The transformed and decoupled form of Equations 5.68 to 5.71 obtained after Daubechies scaling function approximation of the displacements are

$$-I_0\gamma_j^2\widehat{u}_j - A_{11}\frac{d^2\widehat{u}_j}{dx^2} + B_{11}\frac{d^2\widehat{\phi}_j}{dx^2} - A_{13}\frac{d\widehat{\psi}_j}{dx} = 0 \qquad (5.78)$$

$$-I_2\gamma_j^2\widehat{\psi}_j + A_{13}\frac{d\widehat{u}_j}{dx} - B_{13}\frac{d\widehat{\phi}_j}{dx} + A_{33}\widehat{\psi}_j$$
$$-B_{55}\left(\frac{d^2\widehat{w}_j}{dx^2} - \frac{d\widehat{\phi}_j}{dx}\right) - D_{55}\frac{d^2\widehat{\psi}_j}{dx^2} = 0 \qquad (5.79)$$

$$-I_0\gamma_j^2\widehat{w}_j - A_{55}\left(\frac{d^2\widehat{w}_j}{dx^2} - \frac{d\widehat{\phi}_j}{dx}\right) - B_{55}\frac{d^2\widehat{\psi}_j}{dx^2} = 0 \qquad (5.80)$$

$$-I_2\gamma_j^2\widehat{\phi}_j - A_{55}\left(\frac{d\widehat{w}_j}{dx} - \widehat{\phi}_j\right) - B_{55}\frac{d\widehat{\psi}_j}{dx} + B_{11}\frac{d^2\widehat{u}_j}{dx^2}$$
$$-D_{11}\frac{d^2\widehat{\phi}_j}{dx^2} + B_{11}\frac{d\widehat{\psi}_j}{dx} = 0 \qquad (5.81)$$

Similarly, the transformed force boundary conditions are

$$A_{11}\frac{d\widehat{u}_j}{dx} - B_{11}\frac{d\widehat{\phi}_j}{dx} + A_{13}\widehat{\psi}_j = \widehat{P}_j \quad j = 0,\, 1,\dots, n-1 \qquad (5.82)$$

$$A_{55}\frac{d\widehat{w}_j}{dx} + B_{55}\frac{d\widehat{\psi}}{\partial x} - A_{55}\widehat{\phi}_j = \widehat{V}_j \qquad (5.83)$$

$$-B_{11}\frac{d\widehat{u}_j}{dx} + D_{11}\frac{d\widehat{\phi}_j}{dx} - B_{13}\widehat{\psi}_j = \widehat{M}_j \qquad (5.84)$$

$$B_{55}\frac{d\widehat{w}_j}{dx} + D_{55}\frac{d\widehat{\psi}_j}{dx} - B_{55}\widehat{\phi}_j = \widehat{Q}_j \quad j = 0,\, 1,\dots, n-1 \qquad (5.85)$$

The dofs associated with the element formulation are shown in Figure 5.10(b). The element has four degrees of freedom per node, which are $\widehat{u}_j$, $\widehat{w}_j$, $\widehat{\phi}_j$, and $\widehat{\psi}_j$. The reduced ODEs given by Equations 5.78 to 5.81 for composite beam are required to be solved for $\widehat{u}_j$, $\widehat{w}_j$, $\widehat{\phi}_j$, $\widehat{\psi}_j$ and the actual solutions $u(x,t)$, $w(x,t)$, $\phi(x,t)$, $\psi(x,t)$ are obtained using inverse transform.

The formulation of WSFE for such higher order composite beam is similar to that for Euler-Bernoulli beam except the higher complexities associated with the former structure. Here, the WSFE formulation can be started by considering a displacement vector as follows,

$$\{\widehat{v}\} = \begin{Bmatrix} \widehat{v}_1 \\ \widehat{v}_2 \\ \widehat{v}_3 \\ \widehat{v}_4 \end{Bmatrix} = \begin{Bmatrix} \widehat{u}(x) \\ \widehat{w}(x) \\ \widehat{\phi}(x) \\ \widehat{\psi}(x) \end{Bmatrix} \tag{5.86}$$

Now, assuming a solution of the displacements $\widehat{\mathbf{v}}$,

$$\{\widehat{v}\} = \{\widehat{v}_0\}e^{-\imath kx} \qquad \{\widehat{v}_0\} = \begin{Bmatrix} \widehat{u}_0 \\ \widehat{w}_0 \\ \widehat{\phi}_0 \\ \widehat{\psi}_0 \end{Bmatrix} \tag{5.87}$$

Substituting the above solution in the ODEs given by Equations 5.78 to 5.81, a PEP can be posed as

$$\{\mathbf{A}_2 k^2 + \mathbf{A}_1 k + \mathbf{A}_0\}\{\widehat{v}_0\} = 0 \tag{5.88}$$

Here,

$$\mathbf{A_2} = \begin{bmatrix} \gamma^2 I_0 & 0 & 0 & 0 \\ 0 & \gamma^2 I_0 & 0 & 0 \\ 0 & 0 & \gamma^2 I_2 + A_{55} & 0 \\ 0 & 0 & 0 & \gamma^2 I_2 + A_{55} \end{bmatrix}$$

$$\mathbf{A_1} = \begin{bmatrix} 0 & 0 & 0 & \imath A_{13} \\ 0 & 0 & -\imath A_{55} & 0 \\ 0 & \imath A_{55} & 0 & \imath(-B_{13} + B_{55}) \\ -\imath A_{13} & 0 & \imath(B_{13} + B_{55}) & 0 \end{bmatrix}$$

$$\mathbf{A_0} = \begin{bmatrix} A_{11} & 0 & -B_{11} & 0 \\ 0 & A_{55} & 0 & B_{55} \\ -B_{11} & 0 & D_{11} & 0 \\ 0 & B_{55} & 0 & D_{55} \end{bmatrix}$$

The wavenumbers k are obtained as eigenvalues of the above PEP. The eight wavenumbers will be a set of two conjugate wavenumbers, each pair representing the forward and backward moving waves for a mode. The vector $\widehat{\mathbf{v}}_0$ are the eigenvectors of the PEP corresponding to each of the wavenumbers. This solution of the PEP gives these eigenvectors as an 4×8 eigenvector matrix $[R]$ of the form

$$[R] = \begin{bmatrix} R_{11} \cdots & \cdots R_{18} \\ R_{21} \cdots & \cdots R_{28} \\ R_{31} \cdots & \cdots R_{38} \\ R_{41} \cdots & \cdots R_{48} \end{bmatrix} \tag{5.89}$$

Therefore, the solution for $\{\widehat{v}\}$ can be written in matrix form as

$$\{\widehat{v}\} = [R][\Theta]\{a\} \tag{5.90}$$

Here, $[\Theta]$ is a diagonal matrix with the diagonal terms $[e^{-\imath k_1 x}, e^{-\imath k_1(L-x)},$ $e^{-\imath k_2 x}, e^{-\imath k_2(L-x)}, e^{-\imath k_3 x}, e^{-\imath k_3(L-x)}, e^{-\imath k_4 x}, e^{-\imath k_4(L-x)}]$. $\{a\} = \{C_1 \ C_2 \ \ldots C_7 \ C_8\}^T$ are the unknown constants which will be derived from the four boundary conditions at each of the two nodes. Thus, the solution of the reduced ODEs given by Equations 5.78 to 5.81 for each of the displacement can be written as

$$\widehat{v}_m = \sum_{l=1}^{8} R_{ml} C_l e^{-\imath k_l x} \quad \text{for} \quad m = 1 \text{ to } 4 \tag{5.91}$$

Next, imposing the boundary conditions on the displacements as

$$\widehat{u}(x)|_{@x=0} = \widehat{u}_1 = \begin{bmatrix} R_{11} & \cdots & R_{18} \end{bmatrix} [\Theta_1] \{a\}$$
$$= [\mathbf{T}_{11}]\{a\} \tag{5.92}$$

Here, $[\Theta_1]$ is a diagonal matrix with diagonal terms $[1, e^{-\imath k_1 L}, 1, e^{-\imath k_2 L}, 1, e^{-\imath k_3 L}, 1, e^{-\imath k_4 L}]$. Similarly, for other displacements, the displacement boundary conditions can be written as

$$\widehat{w}(x)|_{@x=0} = \widehat{w}_1 = \begin{bmatrix} R_{21} & \cdots & R_{28} \end{bmatrix} [\Theta_1] \{a\}$$
$$= [\mathbf{T}_{12}]\{a\} \tag{5.93}$$
$$\widehat{\phi}(x)|_{@x=0} = \widehat{\phi}_1 = \begin{bmatrix} R_{31} & \cdots & R_{38} \end{bmatrix} [\Theta_1] \{a\}$$
$$= [\mathbf{T}_{13}]\{a\} \tag{5.94}$$
$$\widehat{\psi}(x)|_{@x=0} = \widehat{\psi}_1 = \begin{bmatrix} R_{41} & \cdots & R_{48} \end{bmatrix} [\Theta_1] \{a\}$$
$$= [\mathbf{T}_{14}]\{a\} \tag{5.95}$$

The remaining four displacement boundary conditions at the other node can be written as

$$\widehat{u}(x)|_{@x=L} = \widehat{u}_2 = \begin{bmatrix} R_{11} & \cdots & R_{18} \end{bmatrix} [\Theta_2] \{a\} = [T_{15}]\{a\} \tag{5.96}$$

Here, $[\Theta_2]$ is a diagonal matrix with diagonal terms $[e^{-\imath k_1 L}, 1, e^{-\imath k_2 L}, 1, e^{-\imath k_3 L}, 1, e^{-\imath k_4 L}, 1]$. The other displacements at the node can be written as

$$\widehat{w}(x)|_{@x=L} = \widehat{w}_2 = \begin{bmatrix} R_{21} & \cdots & R_{28} \end{bmatrix} [\Theta_2] \{a\} = [T_{16}]\{a\} \tag{5.97}$$
$$\widehat{\phi}(x)|_{@x=L} = \widehat{\phi}_2 = \begin{bmatrix} R_{31} & \cdots & R_{38} \end{bmatrix} [\Theta_2] \{a\} = [T_{17}]\{a\} \tag{5.98}$$
$$\widehat{\psi}(x)|_{@x=L} = \widehat{\psi}_2 = \begin{bmatrix} R_{41} & \cdots & R_{48} \end{bmatrix} [\Theta_2] \{a\} = [T_{18}]\{a\} \tag{5.99}$$

Combining Equations 5.92 to 5.99, the nodal displacement vector can be written as

$$\{\widehat{\mathbf{u}}^e\} = \begin{Bmatrix} \widehat{u}_1 \\ \widehat{w}_1 \\ \widehat{\phi}_1 \\ \widehat{\psi}_1 \\ \widehat{u}_2 \\ \widehat{w}_2 \\ \widehat{\phi}_2 \\ \widehat{\psi}_2 \end{Bmatrix} = \begin{bmatrix} \mathbf{T}_{11} \\ \mathbf{T}_{12} \\ \vdots \\ \mathbf{T}_{17} \\ \mathbf{T}_{18} \end{bmatrix} \{a\} = [\mathbf{T}_1]\{a\} \tag{5.100}$$

Next, substituting the solutions given by Equation 5.91 in the force boundary conditions given by Equations 5.82 to 5.85, they can be written as

$$\widehat{P}(x)|_{@x=0} = \widehat{P}_1 = [\acute{R}_{11} \cdots \acute{R}_{18}][\boldsymbol{\Theta}_1]\{a\} = [T_{21}]\{a\} \tag{5.101}$$

where $\acute{R}_{1l} = -\imath k_l(A_{11}R_{1l} - B_{11}R_{3l}) + A_{13}R_{4l}$

$$\widehat{V}(x)|_{@x=0} = \widehat{V}_1 = [\acute{R}_{21} \cdots \acute{R}_{28}][\boldsymbol{\Theta}_1]\{a\} = [T_{22}]\{a\} \tag{5.102}$$

where $\acute{R}_{2l} = -\imath k_l(A_{55}R_{2l} - B_{55}R_{4l}) - A_{55}R_{3l}$

$$\widehat{M}(x)|_{@x=0} = \widehat{M}_1 = [\acute{R}_{31} \cdots \acute{R}_{38}][\boldsymbol{\Theta}_1]\{a\} = [T_{23}]\{a\} \tag{5.103}$$

where $\acute{R}_{3l} = -\imath k_l(-B_{11}R_{1l} + D_{11}R_{3l}) - B_{13}R_{4l}$

$$\widehat{Q}(x)|_{@x=0} = \widehat{Q}_1 = [\acute{R}_{41} \cdots \acute{R}_{48}][\boldsymbol{\Theta}_1]\{a\} = [T_{24}]\{a\} \tag{5.104}$$

where $\acute{R}_{4l} = -\imath k_l(B_{55}R_{2l} + D_{55}R_{4l}) - B_{55}R_{3l}$

Similarly, at the other node the forces can be written as

$$\widehat{P}(x)|_{@x=L} = \widehat{P}_2 = [\acute{R}_{11} \cdots \acute{R}_{18}][\boldsymbol{\Theta}_2]\{a\} = [T_{25}]\{a\} \tag{5.105}$$

$$\widehat{V}(x)|_{@x=L} = \widehat{V}_2 = [\acute{R}_{21} \cdots \acute{R}_{28}][\boldsymbol{\Theta}_2]\{a\} = [T_{26}]\{a\} \tag{5.106}$$

$$\widehat{M}(x)|_{@x=L} = \widehat{M}_2 = [\acute{R}_{31} \cdots \acute{R}_{38}][\boldsymbol{\Theta}_2]\{a\} = [T_{27}]\{a\} \tag{5.107}$$

$$\widehat{Q}(x)|_{@x=L} = \widehat{Q}_2 = [\acute{R}_{41} \cdots \acute{R}_{48}][\boldsymbol{\Theta}_2]\{a\} = [T_{28}]\{a\} \tag{5.108}$$

Again, augmenting Equations 5.101 to 5.108, the nodal force vector can be obtained as

$$\{\widehat{\mathbf{F}}^e\} = \begin{Bmatrix} \widehat{P}_1 \\ \widehat{V}_1 \\ \widehat{M}_1 \\ \widehat{Q}_1 \\ \widehat{P}_2 \\ \widehat{V}_2 \\ \widehat{M}_2 \\ \widehat{Q}_2 \end{Bmatrix} = \begin{bmatrix} \mathbf{T}_{21} \\ \mathbf{T}_{22} \\ \vdots \\ \mathbf{T}_{27} \\ \mathbf{T}_{28} \end{bmatrix} \{a\} = [\mathbf{T}_2]\{a\} \tag{5.109}$$

Eliminating the unknown constant vector $\{a\}$ from Equations 5.100 and 5.109, the nodal displacement vector can be related to the nodal force vector as

$$\{\widehat{\mathbf{F}}^e\} = [\mathbf{T}_2][\mathbf{T}_1]^{-1}\{\widehat{\mathbf{u}}^e\} = [\widehat{\mathbf{K}}]\{\widehat{\mathbf{u}}^e\} \tag{5.110}$$

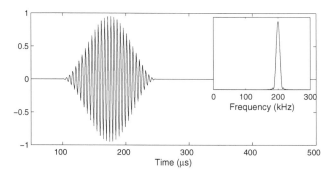

FIGURE 5.11:   Tone-burst signal modulated at 200 kHz and its Fourier transform (inset)

Here, $[\widehat{\mathbf{K}}]$ is the elemental dynamic stiffness matrix. This elemental matrix can be assembled to obtain the global dynamic stiffness matrix of more complex structures, a few examples being, beam with ply-drop, beam with embedded PZT patches as sensors or actuators. Knowing the nodal displacements $\{\widehat{\mathbf{u}}^e\}$ from Equation 5.110, it can be substituted back into Equation 5.100. The constants $\{a\}$ can be obtained as

$$\{a\} = [T_1]^{-1}\{\widehat{u}^e\} \tag{5.111}$$

These constants can be substituted back into Equation 5.90 to obtain the displacements at any arbitrary point on the beam.

5.11   Time domain response of higher order composite beam

The coupled wave propagation in the higher order composite beam simulated using the formulated WSFE is obtained for both broad-band impulse load as shown in Figure 5.2 and narrow-banded modulated sinusoidal pulse or tone-burst signals. An example of tone-burst signal modulated at a frequency of 200 kHz is shown in Figure 5.11. Such tone-burst signals travel non-dispersively and are commonly used for damage detection purposes. The responses are compared with 2-D FE results and also with corresponding FSFE results. All the numerical experiments presented are performed on a AS4/3501-6 graphite-epoxy composite beam with four plies. The beam has a depth of $h = 0.01$ m and width $b = 0.01$ m and the material properties are given in Table 5.2.

TABLE 5.2: Properties of
AS4/3501-6 graphite-epoxy beams

Material Properties:	
E_{11}, Gpa	141.9
E_{22}, Gpa	9.78
$G_{12} = G_{13}$, Gpa	6.13
G_{23}, Gpa	4.80
ν_{12}	0.42
ρ	1449 kg/m^3

5.11.1   Response to impulse load

Wave propagation is studied in a cantilever beam due to the impulse load applied at the free end. In Figure 5.12(a), the tip axial velocities in the undamped beam of length $L = 0.5$ m and ply orientation $[0_2/90_2]$ due the impulse load applied at the tip in axial direction are plotted. The results obtained using WSFE are compared with 1-D FE results and they match very well. In Figure 5.12(b), the tip transverse velocities in the undamped beam of length $L = 0.25$ m and ply orientation $[0_2/90_2]$ due the impulse load applied at the tip in transverse direction are plotted. Even here, the WSFE and 1-D FE results compare well. Here, 1-D FE is used for validation instead of 2-D FE as Euler-Bernoulli beam response cannot be accurately simulated using 2-D FE. Both first order shear deformation theory (FSDT) and Euler-Bernoulli theory (EBT) are considered. The EBT solutions are derived from the formulated FSDT solution by considering $A_{55} = \infty$. WSFE solutions are obtained with $N = 22$ and $\triangle t = 2$ μs with $T_w = 512$ μs. FE solutions are obtained using 2000 1-D beam elements and Newmark's time integration scheme with time step of 1 μs. In addition, it can be seen from Figure 5.12(b) that effect of neglecting transverse shear causes substantial reduction of transverse wave velocity.

In Figures 5.13(a) and (b), the tip axial and transverse velocities in a un-damped cantilever beam due to the tip unit impulse load applied in axial and transverse directions respectively are plotted for different ply orientations. The velocities are simulated using WSFE with $N = 22$, $\triangle t = 2$ μs, and $T_w = 512$ μs. The length of the beam considered is $L = 0.5$ m for measuring the axial velocity and is $L = 0.25$ m for transverse velocity. The three different ply orientations considered are $[0_4]$, $[0_2/60_2]$, and $[0_2/90_2]$. It can be seen from the figures that both the axial and transverse velocities are highest for $[0_4]$ ply layup while least for $[0_2/90_2]$ ply layup. Next, the WSFE solutions are compared with FSFE to emphasize the advantages of WSFE for wave propagation analysis in finite length structures. In Figure 5.14, the tip axial velocities in a $[0_2/60_2]$ cantilever beam with $L = 0.25$ m due to the impulse load applied at tip in transverse direction are plotted. Though WSFE can be used effectively for analysis of undamped finite length structures, for compar-

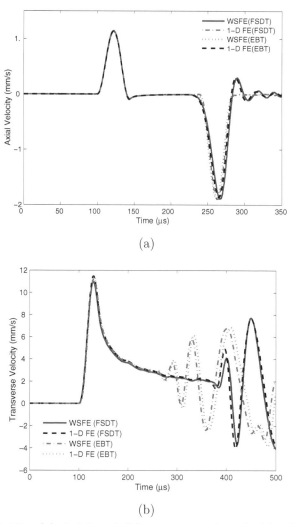

(a)

(b)

FIGURE 5.12:   (a) Axial and (b) transverse tip velocities in a $[0_2/90_2]$ graphite-epoxy beam due to tip impulse load applied in axial and transverse directions respectively

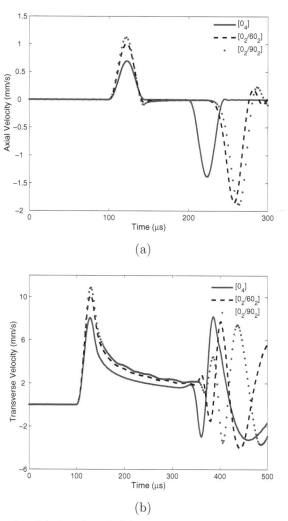

FIGURE 5.13:   (a) Axial and (b) transverse tip velocities in a graphite-epoxy beam with different ply orientations due to tip impulse load applied in axial and transverse directions respectively

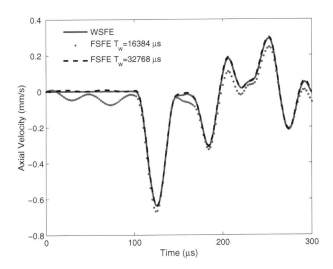

FIGURE 5.14: Axial tip velocities in a graphite-epoxy $[0_2/60_2]$ beam due to tip impulse load applied in transverse direction

ison with FSFE, which does not work for such cases, a damping of $\eta = 0.1$ is considered. For both FSFE and WSFE is $\Delta t = 2$ μs, WSFE results, however, are obtained with $T_w = 512$ μs while FSFE requires much higher T_w to remove the distortions due to wrap around problem. It can be seen from Figure 5.14 that FSFE solution with $T_w = 16384$ μs shows large distortions which decrease by increasing T_w further to 32768 μs.

5.11.2 Response to modulated pulse load

To study the presence of different propagating coupled modes in an asymmetric beam, the responses to narrow banded sinusoidal pulse modulated at a high frequency are simulated using WSFE. For such loading the waves propagate non-dispersively. The load is applied at a point **C** on an infinite beam as shown in Figure 5.15 and the velocities are measured at point **D** at a distance L from **C**. Here, a ply orientation of $[0_2/60_2]$ and $L = 2.0$ m is considered. The load is modulated at 200 kHz, such that the loading frequency is above the highest cut-off frequency which is approximately equal to 165 kHz for contractional mode for this ply layup [56]. In Figures 5.16(a) and (b), the axial and transverse velocities at **C** due to the pulse applied at **D** in transverse direction are plotted. It shows the presence of all the four propagating modes, namely, axial, transverse, shear, and contraction. Similarly, Figures 5.17(a) and (b) show the axial and transverse velocities at **C** due to the pulse applied at **D** in axial direction. In all the above experiments, simulations are done with $N = 22$ and $\Delta t = 1$ μs.

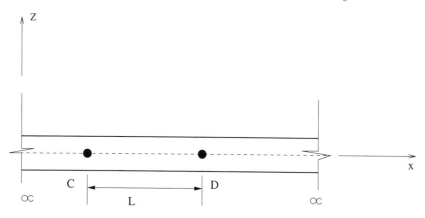

FIGURE 5.15: An infinite beam to observe non-dispersive wave modes

In summary, the chapter presents the wavelet-based spectral finite element method for time domain analysis of wave propagation in one-dimensional structural waveguides. The waveguides studied here as examples are isotropic rod under axial deformation, isotropic Euler-Bernoulli beam, two-dimensional frame, and higher order composite beam. The wavelet-based method presented here can simulate wave response of finite dimensional structures unlike the existing Fourier transform-based methods and this can be stated as a major advancement in the area of numerical modeling for wave propagation analysis. In the following chapter, the WSFE method developed here will be implemented to study the frequency dependent wave characteristics like spectrum and dispersion relations.

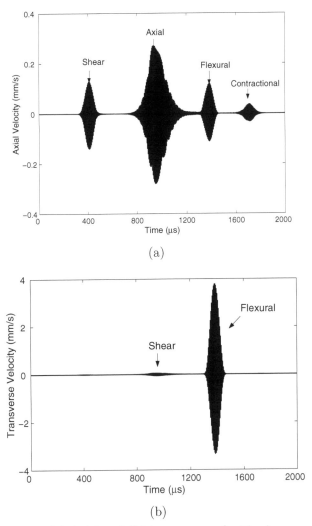

(a)

(b)

FIGURE 5.16:   (a) Axial and (b) transverse velocities in a graphite-epoxy infinite $[0_2/60_2]$ beam due to modulated sinusoidal pulse at 200 kHz applied in transverse direction

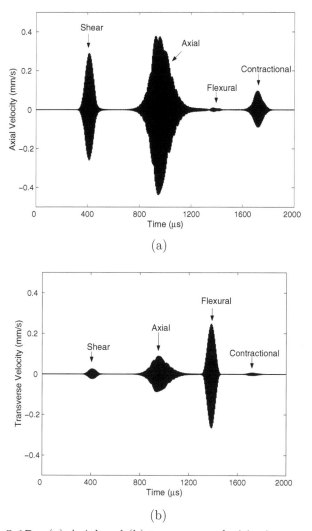

FIGURE 5.17: (a) Axial and (b) transverse velocities in a graphite-epoxy infinite $[0_2/60_2]$ beam due to modulated sinusoidal pulse at 200 kHz applied in axial direction

Chapter 6

Wavelet Spectral Finite Element: Frequency Domain Analysis

An important advantage of spectral method is that it allows simultaneous time and frequency domain analysis. Study of certain frequency dependent wave characteristics like wavenumbers and wave speeds are essential to understand the wave behavior. These frequency dependent wave properties can also be extracted from WSFE formulation but only up to a certain fraction of the Nyquist frequency [63]. This constraint results from the loss in frequency resolution due to the improvement in time resolution in wavelet analysis, where the basis functions are bounded both in time and frequency. A price has to be paid in frequency domain in order to obtain a bound in the time domain. In this chapter, the derivation of wavenumbers and wave speeds from WSFE formulation is presented. The consequence of this analysis is to impose a constraint on the time sampling rate for the simulation with WSFE, to avoid spurious dispersion.

6.1 Frequency domain analysis: periodic boundary condition

In Chapter 5, we studied the non-periodic WSFE formulation for time domain analysis of wave propagation. There, the boundaries are treated using wavelet extrapolation technique. Apart from such non-periodic formulation, WSFE can also be formulated assuming periodicity of the solution. Though the latter method is not capable of accurately simulating the time domain wave response, it helps in deriving the frequency domain wave properties. In this approach, the function $u(x,t)$ is assumed to be periodic in time. Considering the discretized $u(x,\tau)$ to be periodic with time period t_f in the Equation 5.8, the unknown coefficients u_j on LHS are taken as

$$u_{-1} = u_{n-1}$$
$$u_{-2} = u_{n-2}$$
$$\vdots \qquad \vdots$$

117

$$u_{-N+2} = u_{n-N+2}$$

$$(6.1)$$

Similarly the unknown coefficients on RHS, i.e., u_n, $u_{n+1}, \ldots, u_{n+N-2}$ are equal to u_0, $u_1, \ldots, u_{N-2}$ respectively. With the above assumption, the coupled ODEs given by Equation 5.8 can be written in matrix form as

$$\left\{ \frac{d^2 u_j}{dx^2} \right\} = \left(\frac{\eta A}{EA} \Lambda^1 + \frac{\rho A}{EA} \Lambda^2 \right) \{u_j\} \qquad (6.2)$$

where Λ^1 and Λ^2 are $n \times n$ circulant connection coefficient matrices and have the form

$$\Lambda^1 = \frac{1}{\Delta t} \begin{bmatrix} \Omega_0^1 & \Omega_{-1}^1 & \cdots & \Omega_{-N+2}^1 & \cdots & \Omega_{N-2}^1 & \cdots & \Omega_1^1 \\ \Omega_1^1 & \Omega_0^1 & \cdots & \Omega_{-N+3}^1 & \cdots & 0 & \cdots & \Omega_2^1 \\ \vdots & \vdots & \cdots & \vdots & \cdots & \vdots & \cdots & \vdots \\ \Omega_{-1}^1 & \Omega_{-2}^1 & \cdots & 0 & \cdots & \Omega_{N-3}^1 & \cdots & \Omega_0^1 \end{bmatrix} \qquad (6.3)$$

Λ^2 for second order derivative has a similar form. For a circulant matrix Λ^1 [21], the eigenvalues α_j are

$$\alpha_j = \sum_{k=-N+2}^{N-2} \Omega_k^1 e^{-2\pi ijk/n} \qquad j = 0,\ 1,\ldots, n-1 \qquad (6.4)$$

and the corresponding orthonormal eigenvectors v_j, $j = 0,\ 1,\ldots, n-1$ are

$$(v_j)_k = \frac{1}{\sqrt{n}} e^{-2\pi ijk/n}, \qquad k = 0,\ 1,\ldots, n-1 \qquad (6.5)$$

For Λ^1, $\Omega_p^1 = -\Omega_{-p}^1$ for $p = 1,\ 2,\ldots, N-2$ and $\Omega_0^1 = 0$ and we can write $\alpha_j = i\lambda_j$ where

$$\lambda_j = -\frac{2}{\Delta t} \sum_{k=1}^{N-2} \Omega_k^1 \sin\left[\frac{2\pi kj}{n} \right] \qquad j = 0,\ 1,\ldots, n-1 \qquad (6.6)$$

Again here, though the second order connection coefficient matrix Λ^2 can be evaluated independently [8], they can also be written as

$$\Lambda^2 = \left[\Lambda^1\right]^2 \qquad (6.7)$$

Thus, Equation 6.2 can be written as

$$\left\{ \frac{d^2 u_j}{dx^2} \right\} = \left(\frac{\eta A}{EA} [\Lambda^1] + \frac{\rho A}{EA} [\Lambda^1]^2 \right) \{u_j\} \qquad (6.8)$$

As discussed earlier, the spectral element formulation involves eigenvalue analysis. This is done to diagonalize the matrix in Equation 6.2 and decouple the

ODEs. For periodic boundary conditions, these eigenvalues α_j and eigenvectors $(v_j)_k$ are known analytically and this decreases the computational cost. Thus, the diagonalized form of Equation 6.8 is

$$\frac{d^2\widehat{u}_j}{dx^2} = \left(\frac{\eta A}{EA}\lambda_j + \frac{\rho A}{EA}\lambda_j^2\right)\widehat{u}_j \qquad j = 0,\,1,\ldots,n-1 \qquad (6.9)$$

where

$$\widehat{u}_j = \Phi^{-1}u_j \qquad (6.10)$$

Φ being the eigenvector matrix. Neglecting damping, Equation 6.9 can be written as

$$\frac{d^2\widehat{u}_j}{dx^2} = -\frac{\rho A}{EA}\lambda_j^2\widehat{u}_j \qquad j = 0,\,1,\ldots,n-1 \qquad (6.11)$$

For periodic solution, the wavelet transformation given by Equation 5.3 can be written as matrix equation [2]

$$\begin{bmatrix} U_0 \\ U_1 \\ U_2 \\ \vdots \\ \vdots \\ \vdots \\ U_{n-1} \end{bmatrix} = \begin{bmatrix} 0 & 0 & 0 & \cdots & \varphi_{N-2} & \cdots & \varphi_2 & \varphi_1 \\ \varphi_1 & 0 & 0 & \cdots & 0 & \cdots & \varphi_3 & \varphi_2 \\ \varphi_2 & \varphi_1 & 0 & \cdots & 0 & \cdots & \varphi_4 & \varphi_3 \\ \vdots & \vdots & \vdots & \cdots & \vdots & \cdots & \vdots & \vdots \\ \varphi_{N-2} & \varphi_{N-3} & \varphi_{N-4} & \cdots & \cdots & \cdots & 0 & 0 \\ \vdots & \vdots & \vdots & \cdots & \vdots & \cdots & \vdots & \vdots \\ 0 & 0 & 0 & \cdots & \varphi_{N-3} & \cdots & \varphi_1 & 0 \end{bmatrix} \begin{bmatrix} u_0 \\ u_1 \\ u_2 \\ \vdots \\ \vdots \\ \vdots \\ u_{n-1} \end{bmatrix} \qquad (6.12)$$

where U_j, φ_j are the values of $u(x,\tau)$ and $\varphi(\tau)$ at $\tau = j$. For such circulant matrix the Equation 6.12 can be replaced by a convolution relation which can be written as

$$\{\widetilde{U}_j\} = \{\widetilde{K}_{\varphi j}.\widetilde{u}_j\} \qquad (6.13)$$

where $\{\widetilde{U}_j\}$, $\{\widetilde{u}_j\}$ are FFT of $\{U_j\}$ and $\{u_j\}$ respectively. $\{\widetilde{K}_{\varphi j}\}$ is FFT of $\{K_\varphi\} = \{0\ \varphi_1\ \varphi_2\ \ldots\ \varphi_{N-2}\ \ldots\ 0\}$ which is the first column of the scaling function matrix given in Equation 6.12. Similarly in Equation 6.8, the matrix Λ^1 is also a circulant matrix and thus it can be written as (neglecting the damping)

$$\left\{\frac{d^2\widetilde{u}_j}{dx^2}\right\} = \frac{\rho A}{EA}\{\widetilde{K}_{\Omega j}^2.\widetilde{u}_j\} \qquad (6.14)$$

where $\{\widetilde{K}_{\Omega j}\}$ are the FFT of $K_\Omega = \{\Omega_0^1\ \Omega_{-1}^1\ \ldots\ \Omega_{-N+2}^1\ \ldots\ \Omega_{N-2}^1\ \ldots\ \Omega_1^1\}$, which is the first column of the connection coefficient matrix Λ^1. Substituting Equation 6.13 in Equation 6.14 we get

$$\left\{\frac{d^2(\widetilde{U}_j/\widetilde{K}_{\varphi j})}{dx^2}\right\} = \frac{\rho A}{EA}\{\widetilde{K}_{\Omega j}^2.(\widetilde{U}_j/\widetilde{K}_{\varphi j})\} \qquad (6.15)$$

or

$$\left\{ \frac{d^2\widetilde{U}_j}{dx^2} \right\} = \frac{\rho A}{EA} \{\widetilde{K}_{\Omega j}^2 . \widetilde{U}_j\} \tag{6.16}$$

It can be easily seen that the FFT coefficients $\widetilde{K}_{\Omega j}$ are equal to the eigenvalues $i\lambda_j$ of the matrix Λ^1 given by Equation 6.6. Thus Equation 6.16 can be written as

$$\frac{d^2\widetilde{U}_j}{dx^2} = -\frac{\rho A}{EA}\lambda_j{}^2\widetilde{U}_j, \qquad j = 0, \ 1, \ldots, n-1 \tag{6.17}$$

It should be mentioned here that, by relating the Equations 6.11 and 6.17, it can be observed that the transformation given by Equation 6.10 is similar to DFT for periodic WSFE formulation.

In Fourier transform-based spectral analysis, the transformed ODEs are of the form

$$\frac{d^2\widetilde{U}_j}{dx^2} = -\frac{\rho A}{EA}\omega_j{}^2\widetilde{U}_j, \qquad j = 0, \ 1, \ldots, n-1 \tag{6.18}$$

where

$$\omega_j = \frac{2\pi j}{n\triangle t} \tag{6.19}$$

It can be seen that for a given sampling rate $\triangle t$, λ_j exactly matches ω_j up to a certain fraction p_N of Nyquist frequency $f_{nyq} = \frac{1}{2\triangle t}$. Thus similar to Fourier transform-based spectral analysis, WSFE can be used directly for studying frequency dependent characteristics like spectrum and dispersion relations but up to a certain fraction of f_{nyq}. p_N is dependent on the order of basis and is more for higher order basis. In Figure 6.1(a), ω_j and λ_j are compared with respect to a fraction of f_{nyq}. This study also helps to determine the sampling rate required depending on the frequency content of excitation loads and order of bases. This has been explained with numerical experiments later.

Unlike λ_j which are real, the eigenvalues γ_j in non-periodic solution shown in Chapter 5 are complex. However, from numerical experiments it is seen that the real part of γ_j matches λ_j which are compared for different order of basis in Figure 6.1(a). The additional imaginary part of γ_j is plotted for different bases in Figure 6.1(b).

6.2 Computation of wavenumbers and wave speeds

As observed from the above mathematical derivation, the parameter λ_j associated with periodic WSFE formulation matches with the frequency ω_j up to a fraction p_N of the Nyquist frequency f_{nyq} depending on the order N of the Daubechies scaling function used. In WSFE, the wavenumbers k and correspondingly the wave speed $\frac{d\omega}{dk}$ is a function of λ_j and the elastic and

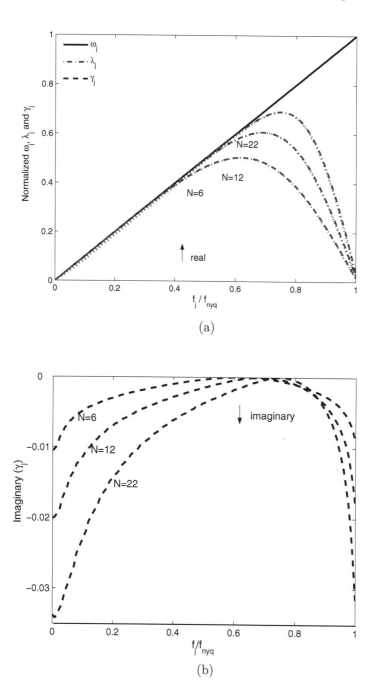

FIGURE 6.1: Comparison of ω_j, λ_j, and γ_j for different order (N) of basis, (a) real part and (b) imaginary part of γ_j

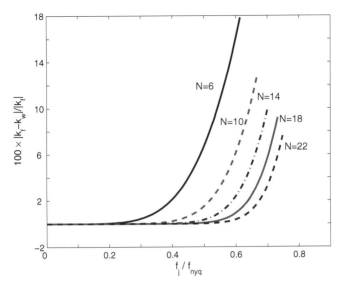

FIGURE 6.2: Comparison of wavenumbers k_f (exact) and k_w (WSFE), for different order (N) of basis

cross-sectional properties of the structure. Thus, for any arbitrary structure, k will be exact up to the frequency $f_N = p_N f_{nyq}$. In Figure 6.2, the percentage error in form of $\frac{|k_f - k_w|}{|k_f|}$ is plotted with respect to the fraction f_j/f_{nyq} for different order of basis functions where k_w are wavenumbers obtained using WSFE and k_f are the actual wavenumbers obtained from Fourier transform respectively. It can be seen that the calculated error is negligible and almost equal to zero up to a certain f_j/f_{nyq} and then increases quite steeply. For a given N, this fraction p_N as defined earlier is the highest frequency fraction up to which the error remains negligible and this can be derived numerically from Figure 6.2 for different N. Though, here the wavenumbers correspond to an elementary rod, the same relation will hold for wavenumbers of other waveguides.

Here numerical examples of spectrum relation, i.e., wavenumber versus frequency plot and dispersion relation, i.e., wave speed versus frequency plot obtained using WSFE, are presented and compared with the exact values of these parameters. All the numerical examples presented here are for an aluminum rod and beam of width $(= 2b)$ 6 mm and depth $(= 2h)$ 25 mm. The elastic properties are as follows, Young's modulus $E = 70$ GPa, shear modulus $G = 27$ GPa, and density $\rho = 2700$ kg/m^3. The shear correction factor $K = 0.85$. As mentioned earlier, the Euler-Bernoulli solutions are obtained considering $GAK = \infty$ and $\rho I = 0$.

6.2.1   Elementary rod

The spectrum relations for an elementary rod are obtained using WSFE with $\triangle t = 1$ μs ($f_{nyq} = 500$ kHz) and $\triangle t = 2$ μs ($f_{nyq} = 250$ kHz) and are presented in Figures 6.3(a) and (b) respectively. The comparison with exact results in each case shows that the non-dimensional wavenumber $k_1 h$ obtained using periodic WSFE is exact up to p_N of f_{nyq} beyond which spurious dispersion is observed. It can be seen from Figures 6.3(a) and (b) that the above mentioned fraction p_N varies only with order of basis N and is independent of the problem. For $N = 22$, $p_N \approx 0.6$, while for $N = 6$, $p_N \approx 0.36$. Thus, for $\triangle t = 1$ μs, wave characteristics can be obtained for a frequency range of $f_{nyq} = 500$ kHz from conventional spectral analysis and for a frequency range of $f_N = 0.6 f_{nyq} = 300$ kHz using WSFE with $N = 22$.

In Figures 6.4(a) and (b), the spectrum relations derived from non-periodic WSFE with $N = 22$ and $N = 6$ and the exact solutions are presented for $\triangle t = 4$ μs and $\triangle t = 8$ μs respectively. As discussed in earlier sections and shown in Figure 6.1(b), non-periodic formulation adds an imaginary part to the real wavenumber k_1 obtained from periodic WSFE. The imaginary part is essentially the damping that is introduced in the formulation to circumvent the wrap around problem and this is inherent in the WSFE formulation. This imaginary part is presented in Figure 6.4(c) for $N = 22$ and $N = 6$ obtained with $\triangle t = 4$ μs.

Next, the dispersion relation for rod is plotted. In the rod the waves are non-dispersive and hence the group speed and phase speeds are the same, i.e., the ratio of $C_g/C_0 = 1$, where C_g represents the group speed and C_0 is the phase speed. In Figure 6.5, the dispersion relation derived from WSFE with $N = 22$ is plotted for $\triangle t = 2$ μs or $f_{nyq} = 250$ kHz and compared to the exact solution. In this figure the non-dimensional group speed C_g/C_0 where $C_g = \text{real}(d\omega/dk_1)$ and $C_0 = \sqrt{EA/\rho A}$ is plotted with respect to frequency. Similar to spectrum relation, WSFE predicts the exact speed up to f_N and is approximately equal to 150 kHz here for $N = 22$.

6.2.2   Euler-Bernoulli and Timoshenko beam

Before studying the wave properties of Timoshenko beam, we will analyze these properties for Euler-Bernoulli beam. The Euler-Bernoulli model is derived from the Timoshenko beam model substituting $A_{55} = \infty$ and rotational inertia equal to zero. In Figure 6.6, the spectrum relation for an Euler-Bernoulli beam obtained using periodic WSFE is presented and compared with exact solution. Similar to elementary rod, even here, the wavenumber derived from WSFE predicts the exact wave behavior up to the fraction p_N of f_{nyq} beyond which it shows spurious dispersions. In Figure 6.6 the non-dimensional wavenumbers $k_2 h$ and $k_3 h$ obtained with $\triangle t = 1$ μs or $f_{nyq} = 500$ kHz are plotted. In Figure 6.7(a) the non-dimensional wavenumbers $k_2 h$ and $k_3 h$ derived from non-periodic WSFE for $\triangle t = 4$ μs are plotted.

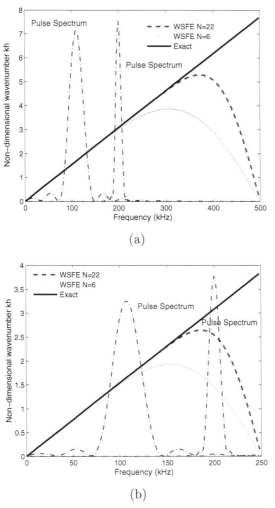

FIGURE 6.3: Spectrum relation for elementary rod $k_1 h$ (and pulse spectrum '$-.-$') for sampling rate (a) $\triangle t = 1 \ \mu$s, (b) $\triangle t = 2 \ \mu$s

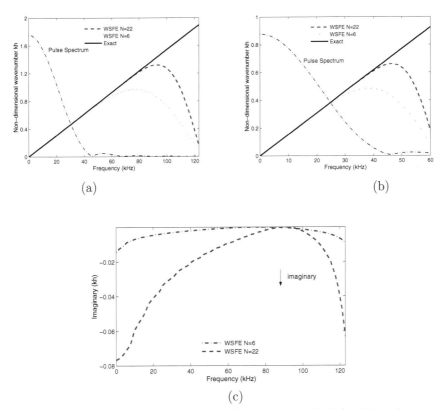

FIGURE 6.4: Spectrum relation for elementary rod k_1h (and impulse spectrum '$-.-$') for sampling rate (a) $\triangle t = 4\ \mu$s, (b) $\triangle t = 8\ \mu$s, and (c) imaginary part of k_1h introduced due to non-periodic WSFE solution with $\triangle t = 4\ \mu$s

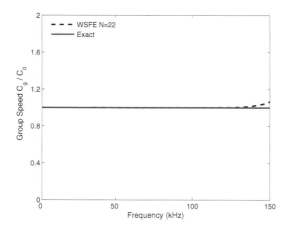

FIGURE 6.5:   Dispersion relation C_g/C_0, $C_0 = \sqrt{EA/\rho A}$ for elementary rod plotted up to $f = 0.6 f_{nyq} = 150$ kHz

In Figure 6.7(b), the imaginary and real part of $k_2 h$ and $k_3 h$ respectively that are introduced due to the imposition of boundary conditions are plotted for $\triangle t = 4$ μs. In Figure 6.8, the dispersion relation of Euler-Bernoulli beam derived from WSFE with $N = 22$ is plotted for $\triangle t = 2$ μs or $f_{nyq} = 250$ kHz and compared with the exact solution. In this figure the non-dimensional group speed C_g/C_0 where $C_g = \text{real}(d\omega/dk_1)$ and $C_0 = \sqrt{EA/\rho A}$ is plotted with respect to frequency. Similar to spectrum relation, WSFE predicts the exact speed up to $f_N \approx 150$ kHz for $N = 22$.

Introduction of shear deformation converts the evanescent mode of elementary beam into a propagating shear mode and this propagation occurs only after certain frequency called cut-off frequency, whose values can be varied by varying material and sectional properties [28]. In Figure 6.9(a), the non-dimensional wavenumbers $k_2 h$ and $k_3 h$ corresponding to flexural and shear modes derived from WSFE ($N = 22$ and $N = 6$) are plotted for $\triangle t = 2$ μs. For the present beam configuration, the cut-off frequency for shear mode k_3 is given by $f_0 = \sqrt{GAK/\rho I} = 62$ kHz and this can be seen from Figure 6.9(a). Thus the shear mode will propagate only for loading with frequency content greater than f_0. Similarly, Figure 6.9(b) shows the spectrum relation for $\triangle t = 8$ μs and up to $f_{nyq} = 64$ μs derived from non-periodic WSFE. The dispersion relations for Timoshenko beam are presented in Figure 6.10. The non-dimensional group speeds C_g/C_0 as described in the earlier sections are derived from WSFE with $N = 22$ and compared with exact solution and $\triangle t = 1$ μs is considered. One of the significant differences between the Euler-Bernoulli and Timoshenko beam is the presence of the second propagating mode. The speeds are plotted up to f_N which is 300 kHz.

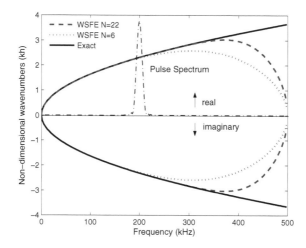

FIGURE 6.6: Spectrum relation for Euler-Bernoulli beam k_2h and k_3h (and pulse spectrum '$-.-$') for sampling rate $\triangle t = 1$ μs

6.2.3   Higher order composite beam

In Figure 6.11, the wavenumbers for a $[0_4]$ AS4/3501-6 graphite-epoxy higher order beam explained in Section 5.9, computed using periodic WSFE (k_w) with $N = 22$, are plotted for all the four modes, i.e., axial, flexural, shear, and contractional. It can be seen that the shear and contractional modes have a cut-off frequency and will propagate for loading with frequencies higher than these cut-off frequencies. Before the cut-off frequency, the wavenumbers corresponding to the shear and contractional modes are imaginary and become real at frequencies higher than the cut-off. For $[0_4]$ ply orientation, these cut-off frequencies for shear and contractional modes are approximately 100 kHz and 162 kHz respectively. For all the modes, comparisons are provided with the exact solution. The sampling rate is $\triangle t = 1$ μs and corresponding Nyquist frequency $f_{nyq} = 500$ kHz.

Next, to study how the wave packets travel at different frequencies, the dispersion relations are plotted for axial, flexural, shear, and contractional modes. In Figure 6.12, the dispersion relation derived from WSFE with $N = 22$ is plotted for $\triangle t = 1$ μs or $f_{nyq} = 500$ kHz. In this figure, the non-dimensional group speeds C_g/C_0 where $C_g =$ real$(d\omega/dk)$ and $C_0 = \sqrt{EA/\rho A}$ are plotted with respect to frequency.

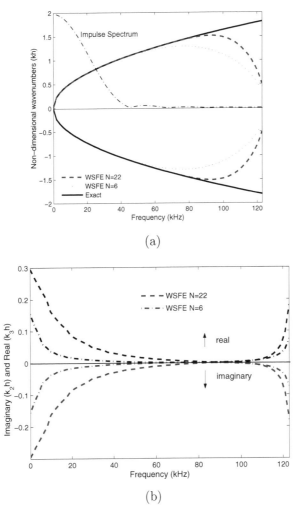

FIGURE 6.7:   (a) Spectrum relation for Euler-Bernoulli beam k_2h and k_3h (and impulse spectrum '$-.-$') for sampling rate $\triangle t = 4 \ \mu$s and (b) imaginary part of k_2h and real part of k_3h introduced due to non-periodic WSFE solution with $\triangle t = 4 \ \mu$s

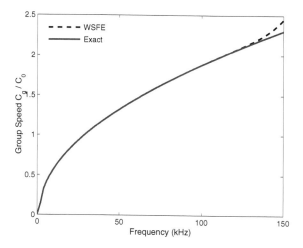

FIGURE 6.8: Dispersion relation C_g/C_0, $C_0 = \sqrt{EA/\rho A}$ for Euler-Bernoulli beam plotted up to $f = 0.6 f_{nyq} = 150$ kHz

6.3 Constraint on time sampling rate

From the above study it has been observed that WSFE is capable of accurate frequency domain analysis similar to the Fourier transform-based analysis of wave propagation, but only up to a fraction of the Nyquist frequency. After this frequency, WSFE shows spurious dispersion. A consequence of the above study is a priori determination of $\triangle t$ required by WSFE depending on the excitation frequency and order of basis N for proper simulation of waves. For a given N, $\triangle t$ should be such that f_N given as $p_N f_{Nyq}$ is greater than the frequency content of the applied load. A larger $\triangle t$ than this leads to spurious dispersion modes of propagation in time domain analysis and this is demonstrated with several examples in this section. First, the response to sinusoidal pulse modulated at 110 kHz and 200 kHz, simulated using non-periodic WSFE is presented. For such loading, the waves propagate non-dispersively even in dispersive media. The pulse spectrum are superimposed in Figures 6.3(a) and (b). The loads are applied at a point **C** on an infinite beam, as shown in Figure 5.15, in axial direction and the axial velocity is measured at **D** at $L = 0.5$ m away from **C**. In Figure 6.13(a), the axial velocities obtained using WSFE (N=22 and N=6) and FE with $\triangle t = 1$ μs for sinusoidal pulse at 200 kHz are presented. It can be seen that in this case, the result obtained using WSFE with $N = 22$ matches exactly with FE, while for $N = 6$, it varies considerably. This observation can be explained using Figure 6.3(a). It can be seen from Figure 6.3(a) that in the frequency range of the applied sinusoidal pulse of

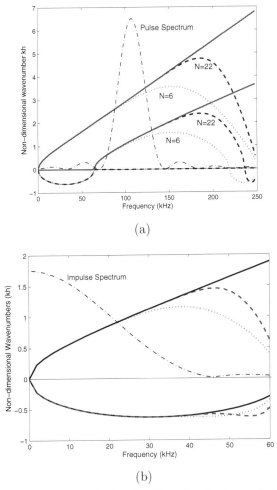

FIGURE 6.9: Spectrum relation for Timoshenko beam k_2h and k_3h using (a) $\triangle t = 2\ \mu s$ (and pulse spectrum '$-.-$'); (b) $\triangle t = 8\ \mu s$ (and impulse spectrum '$-.-$')

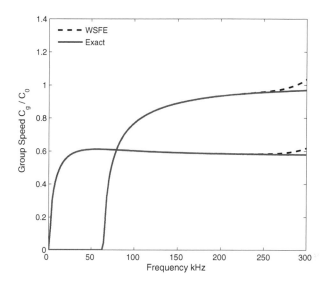

FIGURE 6.10: Dispersion relation C_g/C_0, $C_0 = \sqrt{EA/\rho A}$ for Timoshenko beam

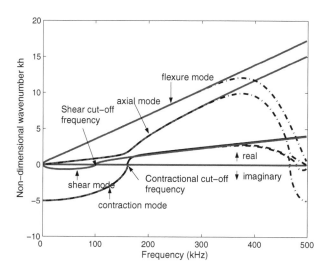

FIGURE 6.11: Spectrum relation for graphite-epoxy $[0_4]$ composite beam

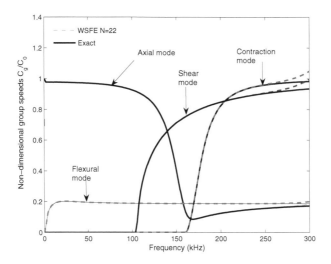

FIGURE 6.12: Dispersion relation C_g/C_0, $C_0 = \sqrt{EA/\rho A}$ for graphite-epoxy $[0_4]$ composite beam plotted up to $f = 0.6 f_{nyq} = 300 \text{kHz}$

200 kHz, WSFE with $N = 22$ predicts exact wavenumbers but WSFE with $N = 6$ fails to do so. Thus the results simulated with $N = 6$ in Figure 6.13(a) cannot capture the correct wave propagation. In Figure 6.13(b) similar axial velocities are presented but for excitation with sinusoidal pulse modulated at 110 kHz. In this case, both the results obtained using WSFE with $N = 22$ and $N = 6$ match exactly with FE results. This is expected as it can be seen from Figure 6.3(a) that for the frequency range of the 110 kHz loading, WSFE with $N = 22$ and $N = 6$ give the correct spectrum relation. From these experiments it can be summarized that the time domain results simulated with WSFE will be accurate if the frequency content of the load is within the range where WSFE can predict the actual spectrum relation accurately. As the allowable range, f_N for a given N is a fixed fraction p_N of f_{nyq}, it can be increased or decreased by decreasing or increasing $\triangle t$ respectively. In Figures 6.14(a) and (b), results similar to Figures 6.13(a) and (b) respectively are presented but here the simulations are done with $\triangle t = 2$ μs. By increasing $\triangle t$, f_{nyq} and correspondingly f_N are decreased to 250 kHz and 150 kHz ($N = 22$), 90 kHz ($N = 6$) respectively. As for $\triangle t = 2$ μs, f_N is less than 200 kHz for both $N = 22$ and $N = 6$, and hence this sampling rate cannot simulate the wave propagation due to 200 kHz modulated sinusoidal loading as shown in Figure 6.14(a). Similarly, for sinusoidal loading of 110 kHz, though WSFE solution with $N = 22$ matches exactly with FE result as shown in Figure 6.14(b), WSFE with $N = 6$ is unable to capture the actual response. This is justified as it can be interpreted from Figure 6.3(b) that for $N = 22$ the excitation frequency lies within the range $f_N = 150$ kHz and is

otherwise for $N = 6$, where $f_N = 90$ kHz. Thus for WSFE with given N, $\triangle t$ or f_{nyq} should be such that f_N is greater than the excitation frequency.

Next, non-periodic WSFE is used for analysis of wave propagation in an undamped fixed free rod of length $L = 0.25$ m. The unit impulse load is applied at the free end in axial direction and axial velocity is measured at the tip. As the excitation frequency here is much lower than that for modulated sinusoidal pulse used in previous examples, higher $\triangle t$ of 4 μs ($f_{nyq} = 128$ kHz) and 8 μs ($f_{nyq} = 64$ kHz) are considered. The impulse spectrum is superimposed in Figures 6.4(a) and (b). For $\triangle t = 4$ μs, all the frequencies, i.e., $f_{nyq} = 128$ kHz, $f_N \approx 77$ kHz for $N = 22$ and $f_N \approx 46$ kHz for $N = 6$, are higher than the excitation frequency and this can also be interpreted from Figure 6.4(a). Thus it is expected that wave propagation analysis with WSFE for $\triangle t = 4$ μs will be correct for both $N = 22$ and $N = 6$. Similarly, as shown in Figure 6.4(b), $f_{nyq} = 64$ kHz, $f_N \approx 38$ kHz for $N = 22$, and $f_N \approx 23$ kHz for $N = 6$. In this case the excitation frequency band exceed, f_N for $N = 6$ and also slightly for $N = 22$ and thus unlike the previous case, this will not simulate the exact response. These predictions are validated with the responses presented in Figures 6.15(a) and (b). In Figure 6.15(a), the tip axial velocities due to tip impulse load obtained using WSFE ($N = 22$ and $N = 6$) and $\triangle t = 4$ μs are plotted. It can be seen that for $\triangle t = 4$ μs WSFE solution matches the exact solution for both $N = 22$ and $N = 6$ as expected. In Figure 6.15(b), similar results are plotted except that $\triangle t = 8$ μs here. It can be observed that WSFE solution for $N = 6$ is highly distorted in this case while for $N = 22$ the response is nearly accurate except some small distortion. This can be justified as f_N for $N = 22$ is very close to the excitation frequency but is otherwise for $N = 6$. Thus $\triangle t$ required for accurate simulation of wave propagation in finite structure due to broadband impulse loading with non-periodic WSFE has to be determined similarly as for the previous cases of modulated sinusoidal excitations. Next, a similar study is performed for Euler-Bernoulli beam and Figure 6.16 shows the transverse velocity of non-dispersive flexural wave due to sinusoidal pulse modulated at 200 kHz applied in transverse direction. The load is applied at a point **C** on the infinite beam shown in Figure 5.15 and the velocity is measured at **D** at a distance $L = 0.5$ m from **C**. In Figure 6.16, the responses obtained with WSFE using $\triangle t = 2$ μs, $N = 22$ and $N = 6$ are compared with exact results. The result for $N = 22$ is accurate whereas for $N = 6$ slight deviations are observed. The explanation for the above observations is similar to that given for elementary rod and is as follows. The spectrum of the pulse is plotted in Figure 6.6 and it can be seen that for $\triangle t = 1$ μs the excitation frequency, i.e., 200 kHz, is slightly more than f_N for $N = 6$ but less for $N = 22$.

Next, WSFE is used to study wave propagation in a cantilever beam with $L = 0.25$ m due to transverse unit impulse load applied at the free end. As like the rod, non-periodic WSFE is used here, for such finite length beam. As the excitation frequency of 44 kHz is lower here, higher $\triangle t$ can be used and the impulse load spectrum is superimposed in spectrum relation shown

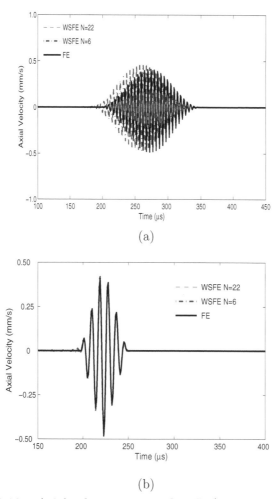

(a)

(b)

FIGURE 6.13:   Axial velocity measured at **D** (Figure 5.15), due to sinusoidal pulse modulated at (a) 200 kHz and (b) 110 kHz, applied at **C** with $\triangle t = 1$ μs and $L = 0.5$m

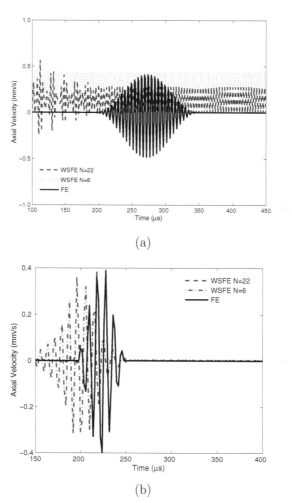

(a)

(b)

FIGURE 6.14: Axial velocity measured at **D** (Figure 5.15), due to sinusoidal pulse modulated at (a) 200 kHz and (b) 110 kHz, applied at **C** with $\triangle t = 2$ μs and $L = 0.5$m

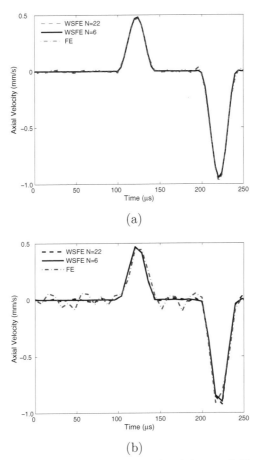

(a)

(b)

FIGURE 6.15: Axial tip velocity of a fixed free rod ($L = 0.25$m) due to tip unit impulse load applied in axial direction with (a) $\triangle t = 4$ μs and (b) $\triangle t = 8$ μs

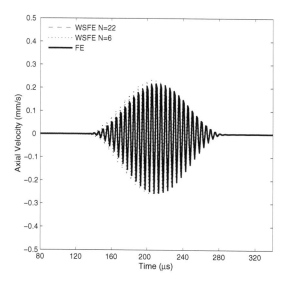

FIGURE 6.16: Transverse velocity measured at **D** (Figure 5.15), due to sinusoidal pulse modulated at 200 kHz applied at **C** with $\triangle t = 1$ μs and $L = 0.5$m

in Figure 6.7(a) with $\triangle t = 4$ μs. For this $\triangle t$, since the excitation frequency is within the range f_N for both $N = 22$ and $N = 6$, WSFE will be able to give the exact results. In Figure 6.17, this tip transverse velocity is plotted for WSFE with $N = 22$, $N = 6$. Finally, the example of the Timoshenko beam with flexural and shear modes is considered. A sinusoidal loading modulated at frequency 110 kHz which is greater than the cut-off frequency is used to capture the simultaneous existence of the coupled modes and the load spectrum is presented in Figure 6.9(a). In Figures 6.18(a) and (b), the responses due to the above load are presented. The transverse velocity is measured at **D** at $L = 2$ m from **C** where the load is applied in the transverse direction. The simulations are done with $\triangle t = 2$ μs and as discussed for elementary rod and Euler-Bernoulli beam, for this $\triangle t$, $f_N = 150$ kHz for $N = 22$ and $f_N = 90$ kHz for $N = 6$ respectively. Thus WSFE with $N = 22$ will predict the exact velocity history and this is validated with Figure 6.18(a) which shows that the solution matches exactly with the corresponding FE solution. However, the solution obtained from WSFE with $N = 6$ plotted in Figure 6.18(b) shows spurious dispersion as expected. Here no propagating mode is clearly visible.

Next, numerical experiments are performed to study the response of Timoshenko due to broad-band impulse load. The details of the loading have been described earlier. The load is applied at the tip of a fixed-free beam of dimension similar to the Euler-Bernoulli beam and the tip transverse velocity is measured. Unlike for Euler-Bernoulli beam here $\triangle t$ is taken even higher and

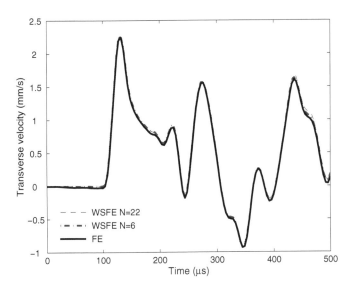

FIGURE 6.17:   Transverse tip velocity of a fixed free Euler-Bernoulli beam ($L = 0.25$m) due to tip unit impulse load applied in transverse direction with $\triangle t = 4$ μs

is equal to 8 μs. In Figure 6.9(b), the superimposed spectrum of the impulse load shows that its frequency content equal to 44 kHz is much greater than $f_N = 23$ kHz for $N = 6$ while slightly for $N = 22$ for which $f_N = 38$ kHz. The time domain response plotted in Figure 6.19 shows that WSFE solution for $N = 22$ matches well with FSFE solution while the solution for $N = 6$ shows deviations.

   To summarize, this chapter explains the use of WSFE formulated and presented in Chapter 4 for frequency domain analysis. It has been derived and seen through examples that WSFE provides the frequency domain wave parameters like spectrum and dispersion relations accurately but only up to a certain fraction of the Nyquist frequency. This analysis also helps to define a threshold on the time sampling rate depending on the frequency content of the load, to avoid spurious dispersion. The following chapters elaborate the extension of WSFE method for modeling two-dimensional waveguides for both time and frequency domain analysis.

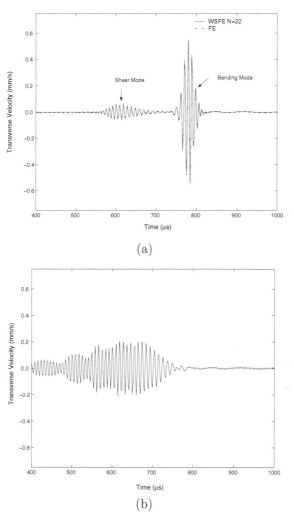

FIGURE 6.18:   Transverse velocity measured at **D** (Figure 5.15), due to sinusoidal pulse modulated at 110 kHz applied at **C** with $\triangle t = 2$ μs and $L = 2.0$m (a) FSFE and WSFE, $N = 22$; (b) WSFE, $N = 6$

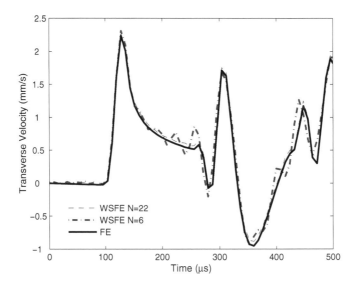

FIGURE 6.19: Transverse tip velocity of a fixed free Timoshenko beam ($L = 0.25$m) due to tip unit impulse load applied in transverse direction with $\triangle t = 8$ μs

Chapter 7

Wavelet Spectral Finite Element: Two-Dimensional Structures

In the last two chapters, the development of WSFE method for one-dimensional structural waveguides has been explained for both time and frequency domain analysis of wave propagation. In this chapter, the extension of the method for modeling of two-dimensional waveguides is discussed. Here, apart from the temporal approximation, where Daubechies scaling functions were used, these functions are again used for approximation in one of the spatial dimensions. This reduces the governing differential partial wave equations to ODEs in terms of one spatial dimensions. These ODEs are solved in a way similar to that explained in Chapter 5. Even here due to the localized nature of the approximation bases, the formulated WSFE can model two-dimensional waveguides with finite dimensions. The method is explained first for isotropic plates and then for axisymmetric cylinders. Next the method is implemented to model structures of higher complexities like folded plate, bi-material cylinder, and anisotropic laminated composites.

7.1 Governing differential wave equations for isotropic plate

The displacement fields, according to Classical Plate Theory (CPT) [73], are

$$U(x, y, z, t) = u(x, y, t) - z \partial w / \partial x \tag{7.1}$$

$$V(x, y, z, t) = v(x, y, t) - z \partial w / \partial y \tag{7.2}$$

$$W(x, y, z, t) = w(x, y, t) \tag{7.3}$$

where $u(x, y, t)$, $v(x, y, t)$, and $w(x, y, t)$ are the axial and transverse displacements respectively along the mid-plane (see Figure 7.1(a)). The mid-plane of the plate is at $z = 0$. The associated non-zero strains are obtained as

$$\begin{Bmatrix} \epsilon_{xx} \\ \epsilon_{yy} \\ \epsilon_{xy} \end{Bmatrix} = \begin{Bmatrix} \partial u / \partial x \\ \partial v / \partial y \\ \partial u / \partial y + \partial v / \partial x \end{Bmatrix}$$

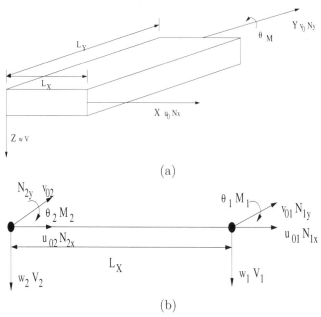

(a)

(b)

FIGURE 7.1: (a) Plate element and (b) nodal forces and displacements

$$+ \left\{ \begin{array}{c} -\partial^2 w/\partial x^2 \\ -\partial^2 w/\partial y^2 \\ -2\partial^2 w/\partial x \partial y \end{array} \right\} = \{\epsilon_0\} + \{\epsilon_1\} \tag{7.4}$$

Here, ϵ_{xx} and ϵ_{yy} are the normal strains in x and y directions respectively, while ϵ_{xy} is the in-plane shear strain. The constitutive relation for isotropic materials is given as

$$\left\{ \begin{array}{c} \sigma_{xx} \\ \sigma_{yy} \\ \sigma_{xy} \end{array} \right\} = \left[\begin{array}{ccc} Q_{11} & Q_{12} & 0 \\ Q_{12} & Q_{22} & 0 \\ 0 & 0 & Q_{66} \end{array} \right] \left\{ \begin{array}{c} \epsilon_{xx} \\ \epsilon_{yy} \\ \epsilon_{xy} \end{array} \right\} \tag{7.5}$$

where σ_{xx} and σ_{yy} are the normal stresses in x and y directions respectively and σ_{xy} is the in-plane shear stress. The expressions for Q_{ij} in terms of Young's modulus E and Poisson's ratio ν are given in reference [73]. The force resultants are defined in terms of these stresses as

$$\left\{ \begin{array}{c} N_{xx} \\ N_{yy} \\ N_{xy} \end{array} \right\} = \int_A \left\{ \begin{array}{c} \sigma_{xx} \\ \sigma_{yy} \\ \sigma_{xy} \end{array} \right\} dA,$$

$$\left\{ \begin{array}{c} M_{xx} \\ M_{yy} \\ M_{xy} \end{array} \right\} = \int_A z \left\{ \begin{array}{c} \sigma_{xx} \\ \sigma_{yy} \\ \sigma_{xy} \end{array} \right\} dA, \tag{7.6}$$

where the integration is performed over the cross-sectional area A. Substituting Equations 7.5 and 7.4 in Equation 7.6, and considering symmetric cross-section, the relations between the force resultants and displacement fields are obtained as

$$
\begin{Bmatrix} N_{xx} \\ N_{yy} \\ N_{xy} \end{Bmatrix} = \begin{bmatrix} A_{11} & A_{12} & 0 \\ A_{12} & A_{22} & 0 \\ 0 & 0 & A_{66} \end{bmatrix} \{\epsilon_0\}
$$

$$
\begin{Bmatrix} M_{xx} \\ M_{yy} \\ M_{xy} \end{Bmatrix} = \begin{bmatrix} D_{11} & D_{12} & 0 \\ D_{12} & D_{22} & 0 \\ 0 & 0 & D_{66} \end{bmatrix} \{\epsilon_1\} \tag{7.7}
$$

The stiffness coefficients A_{ij} and D_{ij} are defined as

$$
[A_{ij}, \ D_{ij}] = \int_A Q_{ij}[1, \ z^2]dA
$$

Similarly, the inertial coefficients used in later part of the section are defined as

$$
[I_0, \ I_2] = \int_A \rho[1, \ z^2]dA
$$

where ρ is the mass density. Total strain Π and kinetic energies T are calculated as

$$
\Pi = \frac{1}{2} \int_0^L \int_A (\sigma_{xx}\epsilon_{xx} + \sigma_{yy}\epsilon_{yy} + \sigma_{xy}\epsilon_{xy})dxdA \tag{7.8}
$$

$$
T = \frac{1}{2} \int_0^L \int_A \rho(\dot{u}^2 + \dot{v}^2 + \dot{w}^2)dxdA \tag{7.9}
$$

Using Hamilton's principle, the minimization of the above energies with respect to the three degrees of freedom $(u, \ v, \ w)$ will give three differential equations, which can be written in terms of the resultant forces and moments as

$$
\partial N_{xx}/\partial x + \partial N_{xy}/\partial y = I_0 \ddot{u} \tag{7.10}
$$

$$
\partial N_{xy}/\partial x + \partial N_{yy}/\partial y = I_0 \ddot{v} \tag{7.11}
$$

$$
\partial^2 M_{xx}/\partial x^2 + 2\partial^2 M_{xy}/\partial x \partial y + \partial^2 M_{yy}/\partial y^2
$$
$$
= I_0 \ddot{w} - I_2(\partial^2 \ddot{w}/\partial x^2 + \partial^2 \ddot{w}/\partial y^2) \tag{7.12}
$$

The governing differential equations can be written in terms of displacements by substituting Equations 7.4 and 7.7 in Equations 7.10 to 7.12 as

$$
A_{11}\partial^2 u/\partial x^2 + (A_{12} + A_{66})\partial^2 v/\partial x \partial y
$$
$$
+ A_{66}\partial^2 v/\partial y^2 = I_0 \ddot{u} \tag{7.13}
$$

$$
A_{66}\partial^2 v/\partial x^2 + (A_{12} + A_{66})\partial^2 u/\partial x \partial y
$$
$$
+ A_{22}\partial^2 v/\partial y^2 = I_0 \ddot{v} \tag{7.14}
$$

$$
D_{11}\partial^4 w/\partial x^4 + 2(D_{12} + 2D_{66})\partial^4 w/\partial x^2 \partial y^2
$$
$$
+ D_{22}\partial^4 w/\partial y^4 = -I_0 \ddot{w}
$$
$$
+ I_2(\partial^2 \ddot{w}/\partial x^2 + \partial^2 \ddot{w}/\partial y^2) \tag{7.15}
$$

The associated boundary conditions are

$$N_x = N_{xx}n_x + N_{xy}n_y, \quad N_y = N_{xy}n_x + N_{yy}n_y \tag{7.16}$$

$$M_x = -M_{xx}n_x - M_{xy}n_y,$$

$$M_y = -M_{xy}n_x - M_{yy}n_y \tag{7.17}$$

$$Q = (\partial M_{xx}/\partial x + \partial M_{xy}/\partial y + I_2\partial\ddot{w}/\partial x)n_x$$

$$+(\partial M_{xy}/\partial x + \partial M_{yy}/\partial y + I_2\partial\ddot{w}/\partial y)n_y \tag{7.18}$$

where N_x and N_y are the normal forces in x and y directions respectively. M_y and M_x are the moments about x and y axes and Q is the transverse shear force in z direction. For edges parallel to the y axis, $n_x = \pm 1$ and $n_y = 0$; thus for modeling a rectangular plate, the boundary conditions given by Equations 7.16 to 7.18 will contain only the terms associated with n_x. In addition, the shear resultant or the Kirchoff shear [28], V is obtained as

$$V = Q - \partial M_{xy}/\partial y \tag{7.19}$$

This modification is done by considering M_{xy} as a couple caused by vertical forces at a small distance apart. This helps to reduce the number of boundary forces to four namely N_x, N_y, V, and M_y as required by CPT. Using the Equation 7.19, the boundary forces for edges parallel to y axis written in terms of displacements are of the form

$$N_x = A_{11}\partial u/\partial x + A_{12}\partial v/\partial y$$

$$N_y = A_{66}(\partial u/\partial y + \partial v/\partial x) \tag{7.20}$$

$$M_y = D_{11}\partial^2 w/\partial x^2 + D_{12}\partial^2 w/\partial y^2 \tag{7.21}$$

$$V = -D_{11}\partial^3 w/\partial x^3 - D_{12}\partial^3 w/\partial x\partial y^2 + I_2\partial\ddot{w}/\partial x \tag{7.22}$$

7.2 Reduction of wave equations through temporal approximation

The formulation of two-dimensional WSFE begins with the reduction of the governing differential equation to another set of PDEs by Daubechies scaling function-based transformation in time. The procedure is very similar to that done for formulating one-dimensional WSFE explained in Chapter 5, where the governing PDEs are reduced to sets of ODEs. The variables are discretized at n points in the time window $[0 \; t_f]$ with $\tau = 0, 1, \ldots, n-1$ being the sampling points as given by Equation 5.2. For example, $u_0(x, y, t)$, the axial mid-plane displacement of plate, can be approximated by scaling function $\varphi(\tau)$ at an arbitrary scale as

$$u(x, y, t) = u(x, y, \tau) = \sum_k u_k(x, y)\varphi(\tau - k), \qquad k \in \mathbf{Z} \tag{7.23}$$

where $u_k(x, y)$ (referred as u_k hereafter) are the approximation coefficients at a certain spatial dimension (x, y). The other displacements can be transformed similarly.

Using the above approximation for $u(x, y, t)$ and similarly for the other displacements $v(x, y, t)$ and $w(x, y, t)$, Equation 7.13 can be written as

$$\sum_k \left(A_{11} \frac{\partial^2 u_k}{\partial x^2} + (A_{12} + A_{66}) \frac{\partial^2 v_k}{\partial x \partial y} + A_{66} \frac{\partial^2 u_k}{\partial y^2} \right) \varphi(\tau - k)$$

$$= \frac{I_0}{\Delta t^2} \sum_k u_k \varphi''(\tau - k) \qquad (7.24)$$

Taking inner product on both sides of Equation 7.24 with the translates of scaling function $\varphi(\tau - j)$, where $j = 0, 1, \ldots, n-1$ and using their orthogonal properties, we get n simultaneous PDEs as

$$A_{11} \frac{\partial^2 u_j}{\partial x^2} + (A_{12} + A_{66}) \frac{\partial^2 v_j}{\partial x \partial y} + A_{66} \frac{\partial^2 u_j}{\partial y^2}$$

$$= \frac{1}{\Delta t^2} \sum_{k=j-N+2}^{j+N-2} \Omega_{j-k}^2 I_0 u_k \qquad j = 0, 1, \ldots, n-1 \qquad (7.25)$$

After treating the boundaries using wavelet extrapolation technique, the above Equations can be written in matrix form as

$$A_{11} \left\{ \frac{\partial^2 u_j}{\partial x^2} \right\} + (A_{12} + A_{66}) \left\{ \frac{\partial^2 v_j}{\partial x \partial y} \right\} + A_{66} \left\{ \frac{\partial^2 u_j}{\partial y^2} \right\}$$

$$= [\Gamma^1]^2 I_0 \{u_j\} \qquad (7.26)$$

Following an eigenvalue analysis of Γ^1, the decoupled PDEs can be obtained as

$$A_{11} \frac{\partial^2 \widehat{u}_j}{\partial x^2} + (A_{12} + A_{66}) \frac{\partial^2 \widehat{v}_j}{\partial x \partial y} + A_{66} \frac{\partial^2 \widehat{u}_j}{\partial y^2} = -I_0 \gamma_j^2 \widehat{u}_j$$

$$j = 0, 1, \ldots, n-1 \qquad (7.27)$$

where $\widehat{u}_j$ and similarly other transformed displacements are

$$\widehat{u}_j = \Phi^{-1} u_j \qquad (7.28)$$

Φ being the eigenvector matrix of Γ^1. Correspondingly, the final transformed forms of Equations 7.14 and 7.15 are

$$A_{66} \frac{\partial^2 \widehat{v}_j}{\partial x^2} + (A_{12} + A_{66}) \frac{\partial^2 \widehat{u}_j}{\partial x \partial y} + A_{22} \frac{\partial^2 \widehat{v}_j}{\partial y^2} = -I_0 \gamma_j^2 \widehat{v}_j \qquad (7.29)$$

$$D_{11} \frac{\partial^4 \widehat{w}_j}{\partial x^4} + 2(D_{12} + 2D_{66}) \frac{\partial^4 \widehat{w}_j}{\partial x^2 \partial y^2} + D_{22} \frac{\partial^4 \widehat{w}_j}{\partial y^4}$$

$$= I_0 \gamma_j^2 \widehat{w}_j - I_2 \gamma_j^2 \left(\frac{\partial^2 \widehat{w}_j}{\partial x^2} + \frac{\partial^2 \widehat{w}_j}{\partial y^2} \right) \qquad (7.30)$$

The force boundary conditions are also transformed in a similar way to the final forms

$$A_{11}\frac{\partial \widehat{u}_j}{\partial x} + A_{12}\frac{\partial \widehat{v}_j}{\partial y} = \widehat{N}_{xj}$$

$$A_{66}\left(\frac{\partial \widehat{u}_j}{\partial y} + \frac{\partial \widehat{v}_j}{\partial x}\right) = \widehat{N}_{yj} \tag{7.31}$$

$$D_{11}\frac{\partial^2 \widehat{w}_j}{\partial x^2} + D_{12}\frac{\partial^2 \widehat{w}_j}{\partial y^2} = \widehat{M}_{yj} \tag{7.32}$$

$$-D_{11}\frac{\partial^3 \widehat{w}_j}{\partial x^3} - D_{12}\frac{\partial^3 \widehat{w}_j}{\partial x \partial y^2} - I_2\gamma_j^2\frac{\partial \widehat{w}_j}{\partial x} = \widehat{V}_j \tag{7.33}$$

$$j = 0,\ 1,\dots,n-1$$

where $\widehat{N}_{xj}$ and similarly $\widehat{N}_{yj}$, $\widehat{M}_{yj}$, $\widehat{V}_j$ are the transformed $N_x(x,y,t)$ and $N_y(x,y,t)$, $M_y(x,y,t)$, $V(x,y,t)$ respectively.

7.3   Reduction of wave equations through spatial approximation

As mentioned earlier, the next step involved is to further reduce each of the transformed and decoupled PDEs to a set of coupled ODEs using Daubechies scaling function approximation in one of the spatial directions. Similar to time approximation, the transformed variable is to be discretized in the spatial window.

Here, the transformed and decoupled PDEs given by Equations 7.27, 7.29, and 7.30 for $j = 0,\ 1,\dots,n-1$ are reduced to a set of coupled ODEs using Daubechies scaling function approximation in Y direction. The transformed axial displacement $\widehat{u}_j$ is discretized at m points in the spatial window $[0,\ L_Y]$, where L_Y is the length in Y direction. Let $\zeta = 0,\ 1,\dots,\ m-1$ be the sampling points, then

$$y = \triangle Y \zeta \tag{7.34}$$

where $\triangle Y$ is the spatial interval between two sampling points. The function $\widehat{u}_j(x,y)$ can be approximated by scaling function $\varphi(\zeta)$ at an arbitrary scale as

$$\widehat{u}_j(x,y) = \widehat{u}_j(x,\zeta) = \sum_k \widehat{u}_{lj}(x)\varphi(\zeta - l), \qquad l \in \mathbf{Z} \tag{7.35}$$

where $\widehat{u}_{lj}(x,y)$ (referred to as $\widehat{u}_{lj}$ hereafter) are the approximation coefficients at a certain spatial dimension x. The other displacements $\widehat{v}_j(x,y)$, $\widehat{w}(x,y)$ can be transformed similarly and Equation 7.27 can be written as

$$A_{11}\sum_l \frac{d^2\widehat{u}_{lj}}{dx^2}\varphi(\zeta - l) + (A_{12} + A_{66})\frac{1}{\triangle Y}\sum_l \frac{d\widehat{v}_{lj}}{dx}\varphi'(\zeta - l)$$

$$+A_{66}\frac{1}{\triangle Y^2}\sum_l \widehat{u}_{lj}\varphi''(\zeta - l) = -I_0\gamma_j^2\sum_l \widehat{u}_{lj}\varphi(\zeta - l) \qquad (7.36)$$

Taking the inner product on both sides of Equation 7.36 with the translates of scaling function $\varphi(\zeta - i)$, where $i = 0, 1, \ldots, m-1$ and using their orthogonal properties, we get m simultaneous ODEs as

$$A_{11}\frac{d^2\widehat{u}_{ij}}{dx^2} + (A_{12} + A_{66})\frac{1}{\triangle Y}\sum_{l=i-N+2}^{i+N-2}\frac{d\widehat{v}_{lj}}{dx}\Omega_{i-l}^1 + A_{66}\frac{1}{\triangle Y^2}\sum_{l=i-N+2}^{i+N-2}\widehat{u}_{lj}\Omega_{i-l}^2$$

$$= -I_0\gamma_j^2\widehat{u}_{ij} \qquad i = 0, 1, \ldots, m-1 \qquad (7.37)$$

where N is the order of Daubechies wavelet and Ω_{i-l}^1 and Ω_{i-l}^2 are the connection coefficients for first and second order derivatives respectively.

It can be seen from the ODEs given by Equation 7.37 that, similar to time approximation, here also certain coefficients $\widehat{u}_{ij}$ near the vicinity of the boundaries ($i = 0$ and $i = m - 1$) lie outside the spatial window $[0 \ L_Y]$ defined by $i = 0, 1, \ldots, m - 1$. These coefficients must be treated properly for finite domain analysis. However here, unlike time approximation, these coefficients are obtained through periodic extension, but only for free lateral edges, while other boundary conditions are imposed quite differently using a restrain matrix [77, 13] and is discussed in detail in the later part of the section.

7.3.1   Un-restraint boundary condition

The unrestrained, i.e., free-free boundary condition, may also be imposed in a similar way using restrain matrix but it has been seen from the numerical experiments that the use of periodic extension gives accurate results. In addition, it allows decoupling of the ODEs using eigenvalue analysis and thus reduces the computational cost. Here, after expressing the unknown coefficients lying outside the finite domain in terms of the inner coefficients considering periodic extension, the ODEs given by Equation 7.37 can be written as a matrix equation of the form

$$A_{11}\left\{\frac{d^2\widehat{u}_{ij}}{dx^2}\right\} + (A_{12} + A_{66})[\Lambda^1]\left\{\frac{d\widehat{v}_{ij}}{dx}\right\} + A_{66}[\Lambda^1]^2\{\widehat{u}_{lj}\}$$

$$= -I_0\gamma_j^2\{\widehat{u}_{ij}\} \qquad (7.38)$$

where $[\Lambda^1]$ is the first order connection coefficient matrix obtained after periodic extension and it is of the form given in Equation 6.3 in Chapter 2, except that $\frac{1}{\triangle t}$ is replaced by $\frac{1}{\triangle Y}$. The coupled ODEs given by Equation 7.38 are decoupled using eigenvalue analysis similar to that done in time approximation as

$$\Lambda^1 = \Psi\Upsilon\Psi^{-1} \qquad (7.39)$$

where Υ is the diagonal eigenvalue matrix and Ψ is the eigenvector matrix of Λ^1. It should be restated here that matrix Λ^1 has a circulant form and its eigen parameters are known analytically [21]. Let the eigenvalues be $\imath\beta_i$, then the decoupled ODEs corresponding to Equation 7.38 are

$$A_{11}\frac{d^2\widetilde{u}_{ij}}{dx^2} - \imath\beta_i(A_{12} + A_{66})\frac{d\widetilde{v}_{ij}}{dx} - \beta_i^2 A_{66}\widetilde{u}_{lj} = -I_0\gamma_j^2\widetilde{u}_{ij}$$
$$i = 0, 1, \ldots, m-1 \tag{7.40}$$

where $\widetilde{u}_j$ and similarly other transformed displacements are

$$\widetilde{u}_j = \Psi^{-1}\widehat{u}_j \tag{7.41}$$

Following exactly the similar steps, the final transformed and decoupled forms of the Equations 7.29 and 7.30 is

$$A_{66}\frac{d^2\widetilde{v}_{ij}}{dx^2} - \imath\beta_i(A_{12} + A_{66})\frac{d\widetilde{u}_{ij}}{dx} - \beta_i^2 A_{22}\widetilde{v}_{ij} = -I_0\gamma_j^2\widetilde{v}_{ij} \tag{7.42}$$

$$D_{11}\frac{d^4\widetilde{w}_{ij}}{dx^4} - 2\beta_i^2(D_{12} + 2D_{66})\frac{d^2\widetilde{w}_{ij}}{dx^2} + \beta_i^4 D_{22}\widetilde{w}_{ij}$$
$$= I_0\gamma_j^2\widetilde{w}_{ij} - I_2\gamma_j^2\left(\frac{d^2\widetilde{w}_{ij}}{dx^2} - \beta_i^2\widetilde{w}_{ij}\right) \tag{7.43}$$

Similarly, the transformed form of the force boundary conditions given by Equations 7.31 to 7.33 are

$$A_{11}\frac{d\widetilde{u}_{ij}}{dx} - \imath\beta_i A_{12}\widetilde{v}_{ij} = \widetilde{N}_{xij} \qquad A_{66}\left(-\imath\beta_i\widetilde{u}_{ij} + \frac{d\widetilde{v}_{0ij}}{dx}\right) = \widetilde{N}_{yij} \tag{7.44}$$

$$D_{11}\frac{d^2\widetilde{w}_{ij}}{dx^2} - \beta_i^2 D_{12}\widetilde{w}_{ij} = \widetilde{M}_{yij} \tag{7.45}$$

$$-D_{11}\frac{d^3\widetilde{w}_{ij}}{dx^3} + \imath\beta_i D_{12}\frac{d^2\widetilde{w}_{ij}}{dx^2} - I_2\gamma_j^2\frac{d\widetilde{w}_{ij}}{dx} = \widetilde{V}_{ij}$$
$$i = 0, 1, \ldots, m-1 \tag{7.46}$$

The final transformed ODEs are given by Equations 7.40, 7.42, 7.43, and the boundary conditions.

7.3.2   Restraint boundary condition

Next, for imposition of other restrained boundary conditions along the lateral edges of plate structure, first Equation 7.37 is written in a different form as

$$A_{11}\frac{d^2\widehat{u}_{ij}}{dx^2} + (A_{12} + A_{66})\frac{1}{\triangle Y}\sum_{l=i-N+2}^{m-1}\frac{d\widehat{v}_{lj}}{dx}\Omega_{i-l}^1$$

$$+ A_{66}\frac{1}{\triangle Y^2}\sum_{l=i-N+2}^{m-1}\widehat{u}_{lj}\Omega_{i-l}^2 = -I_0\gamma_j^2\widehat{u}_{ij} \qquad i = 0, 1, \ldots, m-1 \tag{7.47}$$

This is done by taking the inner product on both sides of Equation 7.36 with the translates of scaling functions $\varphi(\zeta - i)$, where $i = 0$ to $(m-1) - (N-2)$ instead of $i = 0$ to $(m-1)$. Thus, the above Equation 7.47 can be written in a matrix form as

$$A_{11} \left\{ \frac{d^2 \widehat{u}_{ij}}{dx^2} \right\} + (A_{12} + A_{66})[\Lambda_R^1] \left\{ \frac{d\widehat{v}_{ij}}{dx} \right\} + A_{66}[\Lambda_R^1]^2 \{\widehat{u}_{lj}\}$$
$$= -I_0 \gamma_j^2 \{\widehat{u}_{ij}\} \tag{7.48}$$

where $[\Lambda_R^1]$ is a $(m+N-2) \times (m+N-2)$. Now, at the two lateral boundaries given by $y = 0$ and $y = L_Y$, $\widehat{u}_j(x,y)$ or $\widehat{u}_j(x,\zeta)$ can be written in terms of the coefficients $\widehat{u}_{ij}$ as

$$\widehat{u}_j(x,0) = \sum_{l=-N+2}^{0} \widehat{u}_{lj}\varphi(-l) \tag{7.49}$$

$$\widehat{u}_j(x,L_Y) = \widehat{u}_j(x,m-1) = \sum_{l=m-N+2}^{m-1} \widehat{u}_{lj}\varphi(m-1-l) \tag{7.50}$$

From Equations 7.49 and 7.50, the coefficients $\widehat{u}_{(-N+2)j}$ and $\widehat{u}_{(m-1)j}$ are derived in terms of the other coefficients and a $(m+N-2) \times (m+N-2)$ restrain matrix $[R]$ can be formed which is used for transformation from unrestrained to restrained coefficients as

$$\{\widehat{u}_{ij}\}_{restrained} = [R] \{\widehat{u}_{ij}\}_{unrestrained} \tag{7.51}$$

when the boundary conditions are specified in terms of the displacements. Similarly for boundary conditions given in terms of their derivatives, e.g., slope, Equation 7.51 can be written as

$$\left\{ \frac{d\widehat{u}_{ij}}{dy} \right\}_{restrained} = [R] \left\{ \frac{d\widehat{u}_{ij}}{dy} \right\}_{unrestrained} \tag{7.52}$$

The restrain matrix $[R]$ is rank deficient and its order is equal to the number of boundary conditions specified. $[R]$ is formed by inserting two rows obtained from Equation 7.49 and 7.50 to a $(m+N-2) \times m$ identity matrix. Thus, after imposing the restrain, e.g., for a fixed-fixed boundary condition given by $\widehat{u}_j(x,0) = \widehat{v}_j(x,0) = 0$ and $\widehat{u}_j(x,L_Y) = \widehat{v}_j(x,L_Y) = 0$, Equation 7.48 will be of the form

$$A_{11} \left\{ \frac{d^2 \widehat{u}_{ij}}{dx^2} \right\} + (A_{12} + A_{66})[\Lambda_R^1][R] \left\{ \frac{d\widehat{v}_{ij}}{dx} \right\}$$
$$+ A_{66}[\Lambda_R^1]^2[R] \{\widehat{u}_{ij}\} = -I_0 \gamma_j^2 \{\widehat{u}_{ij}\} \tag{7.53}$$

Similarly, the other equation corresponding to Equation 7.29 is

$$A_{66} \left\{ \frac{d^2 \widehat{v}_{ij}}{dx^2} \right\} + (A_{12} + A_{66})[\Lambda_R^1][R] \left\{ \frac{d\widehat{u}_{ij}}{dx} \right\}$$
$$+ A_{22}[\Lambda_R^1]^2[R] \{\widehat{v}_{ij}\} = -I_0 \gamma_j^2 \{\widehat{v}_{ij}\} \tag{7.54}$$

Again, for fixed-fixed boundary condition (out-of-plane loading), $\widehat{w}_j(x, 0) = \partial\widehat{w}_j/\partial y(x, 0) = 0$ and $\widehat{w}_j(x, L_Y) = \partial\widehat{w}_j/\partial y(x, L_Y) = 0$ and the final reduced ODEs obtained from Equation 7.30 are

$$D_{11}\left\{\frac{d^4\widehat{w}_{ij}}{dx^4}\right\} + 2(D_{12} + 2D_{66})[\Lambda_R^1][R][\Lambda_R^1][R]\left\{\frac{d^2\widehat{w}_{ij}}{dx^2}\right\}$$

$$+ D_{22}[\Lambda_R^1]^3[R][\Lambda_R^1][R]\{\widehat{w}_{ij}\} = I_0\gamma_j^2\{\widehat{w}_{ij}\}$$

$$- I_2\gamma_j^2\left(\left\{\frac{d^2\widehat{w}_{ij}}{dx^2}\right\} + [\Lambda_R^1][R][\Lambda_R^1][R]\{\widehat{w}_{ij}\}\right) \qquad (7.55)$$

As mentioned earlier, the matrices involved in Equations 7.53 to 7.55 are rank deficient by two and thus the first and last rows and columns are truncated to solve the equations that are required for the spectral finite element formulation discussed in the next section. The coefficients $\widehat{u}_{(-N+2)j}$ and $\widehat{u}_{(m-1)j}$ are then obtained from the other coefficients using Equations 7.49 and 7.50 respectively. The boundary conditions, Equations 7.31 to 7.33 after transformation and imposition of restrains, are

$$A_{11}\left\{\frac{d\widehat{u}_{ij}}{dx}\right\} + A_{12}[\Lambda_R^1][R]\{\widehat{v}_{ij}\} = \left\{\widehat{N}_{xij}\right\}$$

$$A_{66}\left([\Lambda_R^1][R]\{\widehat{u}_{ij}\} + \left\{\frac{d\widehat{v}_{ij}}{dx}\right\}\right) = \left\{\widehat{N}_{yij}\right\} \qquad (7.56)$$

$$D_{11}\left\{\frac{d^2\widehat{w}_{ij}}{dx^2}\right\} + D_{12}[\Lambda_R^1][R][\Lambda_R^1][R]\{\widehat{w}_{ij}\} = \left\{\widehat{M}_{yij}\right\} \qquad (7.57)$$

$$-D_{11}\left\{\frac{d^3\widehat{w}_{ij}}{dx^3}\right\} - D_{12}[\Lambda_R^1][R]\left\{\frac{d^3\widehat{w}_{ij}}{dx^2}\right\} - I_2\gamma_j^2\left\{\frac{d\widehat{w}_{ij}}{dx}\right\} = \left\{\widehat{V}_{ij}\right\} (7.58)$$

7.4   Wavelet spectral finite element for plate

The degrees of freedom associated with the element formulation are shown in Figure 7.1(b). The element has four degrees of freedom per node, which are $\widetilde{u}_{ij}$, $\widetilde{v}_{ij}$, $\widetilde{w}_{ij}$, and $\partial\widetilde{w}_{ij}/\partial x$. From the Section 7.3, for unrestrained lateral edges, we get a set of decoupled ODEs (Equations 7.40, 7.42, and 7.43) for isotopic plate using CPT, in the transformed wavelet domain. These equations are required to be solved for $\widetilde{u}_{ij}$, $\widetilde{v}_{ij}$, $\widetilde{w}_{ij}$ and the actual solutions $u(x, y, t)$, $v(x, y, t)$, $w(x, y, t)$ are obtained using inverse wavelet transform twice for spatial Y dimension and time. Here, the spectral finite element technique is explained for the decoupled ODEs given by Equations 7.40, 7.42, and 7.43 for unrestrained i.e., free lateral edges. However, for restrained boundary conditions the transformed ODEs given by Equations 7.53 to 7.55 are coupled and spectral finite element formulation for such cases follows similar steps, except that for each time discretization point j, $m - N \times m - N$ matrix ODE is solved instead of m decoupled ODEs.

It can be seen that the transformed decoupled ODEs have forms similar to that in one-dimensional WSFE in Chapter 5 and, thus, WSFE for two-dimensional structure can be formulated following the same method. In this section, the subscripts j and i are dropped hereafter for simplified notations and all the following equations are valid for $j = 0, 1, \ldots, n - 1$ and $i = 0, 1, \ldots, m - 1$ for each j.

The exact interpolating functions for an element of length L_X can be obtained by solving Equations. 7.40, 7.42, and 7.43 after posing them as a PEP as explained in Section 5.10. Let us consider the displacement vector $\tilde{\mathbf{u}}$ as follows,

$$\tilde{\mathbf{u}} = \begin{Bmatrix} \tilde{v}_1 \\ \tilde{v}_2 \\ \tilde{v}_3 \end{Bmatrix} = \begin{Bmatrix} \tilde{u}(x) \\ \tilde{v}(x) \\ \tilde{w}(x) \end{Bmatrix} \tag{7.59}$$

Now, assuming a solution of $\tilde{\mathbf{u}}$ as

$$\tilde{\mathbf{u}} = \tilde{\mathbf{u}}_0 e^{-ikx} \quad \tilde{\mathbf{u}}_0 = \begin{Bmatrix} \tilde{u}_0 \\ \tilde{v}_0 \\ \tilde{w}_0 \end{Bmatrix} \tag{7.60}$$

Substituting Equation 7.60 into the Equations 7.40, 7.42, and 7.43 a PEP of the following form can be posed,

$$\{\mathbf{A}_4 k^4 + \mathbf{A}_2 k^2 + \mathbf{A}_1 k + \mathbf{A}_0\}\tilde{\mathbf{u}}_0 = 0 \tag{7.61}$$

where

$$\mathbf{A}_4 = \begin{bmatrix} 0 & 0 & 0 \\ 0 & 0 & 0 \\ 0 & 0 & D_{11} \end{bmatrix} \quad \mathbf{A}_2 = \begin{bmatrix} -A_{11} & 0 & 0 \\ 0 & -A_{66} & 0 \\ 0 & 0 & 2\beta^2(D_{12} + 2D_{66}) - I_2\gamma^2 \end{bmatrix}$$

$$\mathbf{A}_1 = \begin{bmatrix} 0 & \beta(A_{12} + A_{66}) & 0 \\ \beta(A_{12} + A_{66}) & 0 & 0 \\ 0 & 0 & 0 \end{bmatrix}$$

$$\mathbf{A}_0 = \begin{bmatrix} I_0\gamma^2 & 0 & 0 \\ 0 & I_0\gamma^2 - \beta^2 A_{22} & 0 \\ 0 & 0 & \beta^4 D_{22} - I_0\gamma^2 - I_2\gamma^2\beta^2 \end{bmatrix}$$

As mentioned earlier, the wavenumbers k are obtained as eigenvalues of the PEP given by Equation 7.61. Similarly, the vector $\tilde{\mathbf{u}}_0$ is the eigenvectors corresponding to each of the wavenumbers. The solution of Equation 7.61 gives a 3×8 eigenvector matrix of the form

$$[\mathbf{R}] = \begin{bmatrix} R_{11} \ldots & \ldots R_{18} \\ R_{21} \ldots & \ldots R_{28} \\ R_{31} \ldots & \ldots R_{38} \end{bmatrix} \tag{7.62}$$

Therefore, the solution of $\tilde{\mathbf{u}}$ can be written as

$$\{\tilde{\mathbf{u}}\} = [\mathbf{R}][\Theta]\{a\} \qquad (7.63)$$

where $[\Theta]$ is a diagonal matrix with the diagonal terms $[e^{-k_1 x}, e^{-k_1(L_X - x)}, e^{-k_2 x}, e^{-k_2(L_X - x)}, e^{-k_3 x}, e^{-k_3(L_X - x)}, e^{-k_4 x}, e^{-k_4(L_X - x)}]$.

Here, $\{a\} = \{C_1, C_2, C_3, C_4, C_5, C_6, C_7, C_8\}$ are the unknown coefficients to be determined from transformed nodal displacements $\{\tilde{\mathbf{u}}^e\}$, where $\{\tilde{\mathbf{u}}^e\} = \{\tilde{u}_1 \ \tilde{v}_1 \ \tilde{w}_1 \ \partial \tilde{w}_1/\partial x \ \tilde{u}_2 \ \tilde{v}_2 \ \tilde{w}_2 \ \partial \tilde{w}_2/\partial x\}$ and $\tilde{u}_1 \equiv \tilde{u}(0)$, $\tilde{v}_1 \equiv \tilde{v}(0)$, $\tilde{w}_1 \equiv \tilde{w}(0)$, $\partial \tilde{w}_1/\partial x \equiv \partial \tilde{w}(0)/\partial x$ and $\tilde{u}_2 \equiv \tilde{u}(L_X)$, $\tilde{v}_2 \equiv \tilde{v}(L_X)$, $\tilde{w}_2 \equiv \tilde{w}(L_X)$, $\partial \tilde{w}_2/\partial x \equiv \partial \tilde{w}/\partial x(L_X)$ (see Figure 7.1(b) for the details of degree of freedom the element can support). Thus, similar to the formulation given in Section 5.10 we can relate the nodal displacements and unknown coefficients as

$$\{\tilde{\mathbf{u}}^e\} = [\mathbf{T}_1]\{a\} \qquad (7.64)$$

Similarly, from the forced boundary conditions (Equations 7.44 to 7.46), nodal forces and unknown coefficients can be related as

$$\{\widetilde{\mathbf{F}}^e\} = [\mathbf{T}_2]\{a\} \qquad (7.65)$$

where $\{\widetilde{\mathbf{F}}^e\} = \{\widetilde{N}_{x1} \ \widetilde{N}_{y1} \ \widetilde{V}_1 \ \widetilde{M}_{y1} \ \widetilde{N}_{x2} \ \widetilde{N}_{y2} \ \widetilde{V}_2 \ \widetilde{M}_{y2}\}$ and $\widetilde{N}_{x1} \equiv -\widetilde{N}_x(0)$, $\widetilde{N}_{y1} \equiv -\widetilde{N}_y(0)$, $\widetilde{V}_1 \equiv -\widetilde{V}(0)$, $\widetilde{M}_{y1} \equiv -\widetilde{M}_y(0)$, and $\widetilde{N}_{x2} \equiv \widetilde{N}_x(L_X)$, $\widetilde{N}_{y2} \equiv \widetilde{N}_y(L_X)$, $\widetilde{V}_2 \equiv \widetilde{V}(L_X)$, $\widetilde{M}_{y2} \equiv \widetilde{M}_y(L_X)$ (see Figure 7.1(b)). From Equations 7.64 and 7.65 we can obtain a relation between transformed nodal forces and displacements similar to conventional FE

$$\{\widetilde{\mathbf{F}}^e\} = [\mathbf{T}_2][\mathbf{T}_1]^{-1}\{\tilde{\mathbf{u}}^e\} = [\widetilde{K}^e]\{\tilde{u}^e\} \qquad (7.66)$$

where $[\widetilde{\mathbf{K}}^e]$ is the exact elemental dynamic stiffness matrix. After the constants $\{a\}$ are known from the above equations, they can substituted back to Equation 7.63 to obtain the transformed displacements $\tilde{u}_0$, $\tilde{v}_0$, $\tilde{w}$, $\partial \tilde{w}/\partial x$ at any given x. Following is the Matlab program 7.1 for implementation of two-dimensional WSFE for plate structure for in-plane wave propagation given by Equations 7.13 and 7.14. The input to the program is the stiffness and inertial constants as defined in the equations. These constants can be evaluated using different Matlab functions arbitrarily named in the program as *stiff* and *mass*. The load matrix, again arbitrarily named as *fpulse.unt*, contains the magnitude of the load at a time instance t_i and spatial position y_j.

7.5   Wave propagation in isotropic plates

Here, the formulated two-dimensional WSFE is used to study axial and transverse wave propagations in an isotropic aluminum cantilever plate in both

```
--------------------------------------------------------------
MATLAB PROGRAM 7.1
--------------------------------------------------------------
load fpulse.unt;
F=fpulse; dt=1e-06; % sampling time
[n,m]=size(F); Ly=0.5; dy=Ly/(m-1);
L=2.0; x=2.0; % Length LX
% STIFFNESS & INERTIAL CONSTANTS
[A11,D11,A12,D12,A22,D22,A66,D66]=stiff; I0=mass;
% Daubechies basis function N=DL and order dn
DL=22; d=1;
% Connection Coeffs. for Temporal Approximation
CC2=wavenp(DL,n,d); % Connection Coefficients
%
[VNP,D]=eig(CC2); DNP=j*(1/dt)*diag(D); DL=4;
% Connection Coeffs. for Spatial approximation
K1_omega=(1/dy)*ccoeff(DL,d);
for i1=1:m
    DP(i1)=K1_omega(1,1);
    for q=1:DL-2
        DP(i1)=2*K1_omega(q+1,1)*sin(2*pi*(i1-1)*q/(m))+DP(i1);
    end
end
for i2=1:m,
    for k=1:m
        VP(i2,k)=exp(-2*pi*j*(i2-1)*(k-1)/m);
    end
end
%
FN=inv(VP)*((inv(VNP)*F).');
%
CONST_R=zeros(2,4); CONST_N=zeros(2,4);
% CHARACTERISTIC EQUATIONS
for ii=1:n
    ii
    lambda=DNP(ii);
    for l=1:m
        zeta=DP(l);
    % PEP 1 for axial deformation
    PA1=[-A11 0;0 -A66]; PA2=[0 -zeta*(A12+A66);...
        -zeta*(A12+A66) 0];
    PA3=[lambda^2*I0-zeta^2*A66 0; 0 lambda^2*I0-zeta^2*A22];
    [CONST_R,k_j]=polyeig(PA3,PA2,PA1);
    [kj,I]=sort(k_j);
```

```
------------------------------------------------------------
MATLAB PROGRAM 7.1 (Contd.)

------------------------------------------------------------
      for r=1:4
      CONST_N(:,r)=CONST_R(:,I(r));
      end
%**********************************************************
h=1; for p=1:2 if sign(imag(kj(h)))==1
    TEMP=CONST_N(:,h);
        CONST_N(:,h)=CONST_N(:,h+1);
        CONST_N(:,h+1)=TEMP;
        temp=kj(h);
        kj(h)=kj(h+1);
        kj(h+1)=temp;
    end
    h=h+2;
end
%**********************************************************
EL1=exp(-j*kj(1)*L); EL2=exp(-j*kj(3)*L);
% ESSENTIAL BOUNDARY CONDITION
DG1=diag([1;EL1;1;EL2]); T11=CONST_N*DG1; DG2=diag([EL1;1;EL2;1]);
T12=CONST_N*DG2; T1=[T11;T12];
% FORCED BOUNDARY CONDITION
% axial force
for il=1:4
P(1,il)=-j*kj(il)*A11*CONST_N(1,il)-j*zeta*A12*CONST_N(2,il);
VT(1,il)=-j*A66*(zeta*CONST_N(1,il)+kj(il)*CONST_N(2,il)); end
P1=P*DG1; P2=P*DG2; V1=VT*DG1; V2=VT*DG2;
%
T2=[P1;V1;-P2;-V2]; KD=T2*inv(T1); KD(1:2,:)=[]; KD(:,1:2)=[];
%
U=[FN(1,ii) 0]*inv(KD);
%
ue=[0; 0; U(1,1); U(1,2)]; const=inv(T1)*ue;
DGX=diag([exp(-j*kj(1)*x) exp(-j*kj(1)*(L-x)) exp(-j*kj(3)*x) ...
    exp(-j*kj(3)*(L-x))]);
w_k_t1(1,ii)=-j*lambda*CONST_N(1,:)*DGX*const;
w_k_t2(1,ii)=-j*lambda*CONST_N(2,:)*DGX*const;
    end
end
t=0:dt:(n-1)*dt; WT1=VP*w_k_t1; w_k1=VNP*(WT1.');
v1=imag(w_k1(:,m/2)); % axial wave response @ mid-point
WT2=VP*w_k_t2; w_k2=VNP*(WT2.');
v2=imag(w_k2(:,m/2)); % lateral wave response @ mid-point
```

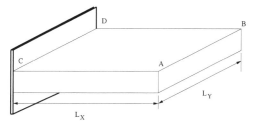

FIGURE 7.2:   Cantilever uniform plate

time and frequency domains. The plate shown in Figure 7.2 is fixed at one edge and free at the other edge along Y-axis. Numerical experiments are performed by considering the other two edges along X-axis to be free-free and fixed-fixed. The dimensions are L_X and L_Y along X- and Y-axes, respectively, while the depth ($= 2h$) is kept fixed at 0.01 m for the uniform plate shown in Figure 7.2. However, both the lengths of the plate are kept small to show the effectiveness of the developed modeling technique in capturing the effects of these edges on the wave propagation behavior. The impulse load is applied at the free edge along the Y-axis and has a spatial distribution of $F(Y) = e^{-(Y/\alpha)^2}$, where α is a constant and can be varied to change the Y-axis variation of the load.

The spatial sampling rate $\triangle Y$ is varied depending on L_Y and load distribution $F(y)$. As mentioned earlier, only one 2-D WSFE is used to simulate the responses of the uniform plate shown in Figure 7.2.

The accuracy of the responses simulated using the developed two-dimensional WSFE is validated with 2-D FE results. The FE meshing is done with 4-noded quadrilateral plane stress elements. Time integration is done using Newmark's scheme with time step 1 μs. The WSFE results are also compared with those obtained using the Fourier transform-based method to emphasize the advantages of the former method for wave propagation analysis of 2-D structures with finite dimension.

7.5.1   Frequency domain analysis

The spectrum relation for the plate with $L_Y = 0.25$ m shown in Figure 7.2 is plotted in Figures 7.3(a) and (b) for axial wave propagation. The real and imaginary parts of the wavenumbers are plotted in Figures 7.3(a) and (b) respectively for a Y wavenumber of 50 with $\triangle t = 8$ μs, i.e., for Nyquist frequency $f_{nyq} = 62.5$ kHz. Comparison is also made with the results obtained using Fourier transform-based method and it can be seen that WSFE predicts accurate wavenumbers, however, up to a certain fraction p_N of the Nyquist frequency f_{nyq}. Here, we see that the wavenumbers have significant real and imaginary parts. That is, the wave, as it propagates, also attenuates. Such waves are called inhomogeneous waves. As said earlier, this fraction p_N depends on the order of the Daubechies scaling function and is ≈ 0.6 for $N = 22$.

First, in Figures 7.4(a) and (b), the real and imaginary parts of the wavenumbers for transverse wave propagation are shown respectively. The wavenumbers are plotted up to the Nyquist frequency $f_{nyq} = 62.5$ kHz for time sampling rate $\triangle t = 8$ μs and the Y-wavenumber considered is 100. From the figures, we see that the waves are inhomogeneous in nature. As in the previous case of axial wave propagation, the wavenumbers obtained from both FSFE and WSFE are presented and can be seen that WSFE gives accurate prediction up to the allowable frequency range, i.e., the fraction p_N of f_{nyq}.

7.5.2   Time domain response of plate to impulse load

Figures 7.5(a) and (b) show the axial velocities of a cantilever plate as in Figure 7.2(a), measured at mid and quarter points respectively, on the free edge AB along the Y-axis. Here, the results simulated with the formulated 2-D WSFE are compared with 2-D FE results for validation. It can be seen that the responses obtained with the two methods match very well. The plate has a finite dimension of $L_X = 4.0$ m and $L_Y = 0.5$ m and is free-free on the other two edges AC and BD along X-axis. L_Y is purposely chosen to be much smaller than L_X so as to show that the developed WSFE can efficiently capture the reflections from lateral edges AC and BD apart of those from the fixed edge CD. The impulse load (Figure 5.2) with $\alpha = 0.05$ for Y-variation is applied along the edge AB in axial direction. As mentioned earlier only one WSFE is used to model the structure and the time window is kept $T_w = 1024$ μs with number of sampling points $n = 512$ and $\triangle t = 2$ μs. The number of discretization points along Y-axis is $m = 64$ and thus the spatial sampling rate is $\triangle Y = L_Y/(m-1) = 0.0079$ m. A very refined mesh with 12864, 4-noded plane stress quadrilateral elements was used for the 2-D FE analysis, while Newmark's scheme with time step 1 μs was used for time integration.

Similar comparison between WSFE and 2-D FE results is made in Figure 7.6(a); however, here the two edges AC and BD are considered fixed. Otherwise the plate dimensions and loading conditions are same as the previous example. The axial velocities plotted are measured at mid-point of AB. Even for this case, the responses compare very well. It can be seen from the figure that the first reflection from fixed edges AC and BD is reversed if compared with the similar response of the free-free plate shown in Figure 7.5(a). In Figure 7.6(b), the axial velocity at the mid-point of AB simulated with a single WSFE is plotted and shows good comparison with 2-D FE result for a plate very similar to that in last example, except $L_Y = 0.25$ m. The unit impulse load is applied along AB in axial direction; however, here $\alpha = 0.03$ for Y-variation. The FE mesh, time integration scheme, and the parameters involved in WSFE modeling are similar to the previous case.

The previous examples considered in-plane wave propagation in an isotropic plate. The next few examples show the implementation and validation of

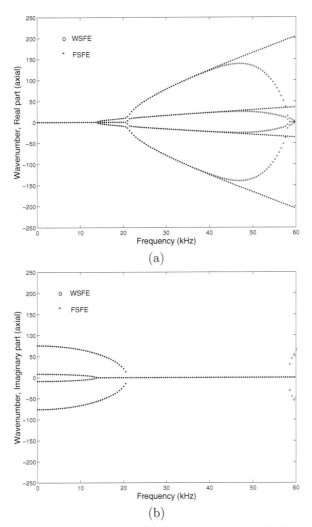

FIGURE 7.3:   The (a) real and (b) imaginary parts of the wavenumbers for axial wave propagation

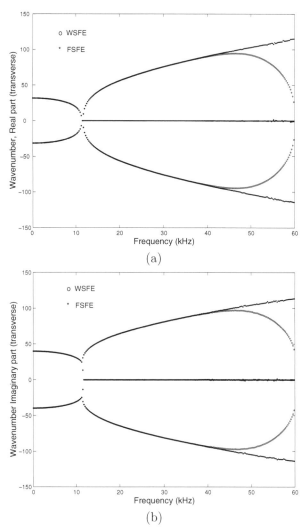

FIGURE 7.4:   The (a) real and (b) imaginary parts of the wavenumbers
for transverse wave propagation

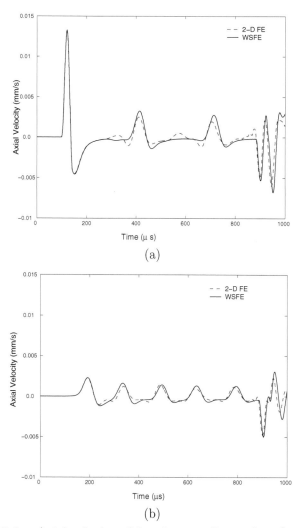

FIGURE 7.5: Axial velocity of free-free cantilever plate (see Figure 7.2) with $L_X = 4.0$ m and $L_Y = 0.5$ m as measured at (a) mid and (b) quarter points of the free end AB

WSFE method for simulation of out-of-plane wave propagation in plate.

The transverse velocities at mid and quarter points on the free edge AB of the cantilever plate shown in Figure 7.2(a) are presented in Figures 7.7(a) and (b) respectively. The responses obtained using the present two-dimensional WSFE method are compared with two-dimensional FE results and a good match is observed. The two edges AC and BD of the plate are free and the dimensions are given as $L_X = 4.0$ m and $L_Y = 0.5$ m. The unit impulse load is applied along the free edge AB in transverse direction and the Y-distribution is obtained using $\alpha = 0.05$. Single WSFE is used to simulate the responses with $n = 1024$, $\triangle t = 2$ μs and thus $T_w = 2048$ μs. The spatial sampling rate is $\triangle Y = 0.0079$ m, number of sampling points being $m = 64$. The FE mesh has 12864, 4-noded plane stress quadrilateral elements and time integration is done using Newmark's scheme with time step 1 μs.

As an advanced problem, the simulation of wave propagation in composite laminates with symmetric and asymmetric ply-up are presented. The formulation of WSFE for such composite structures is not presented mostly because it is a straightforward extension of the formulation for isotropic plates. In addition, formulation of WSFE for higher order composite beam with asymmetric ply lay-up has been presented in Section 5.9 of Chapter 5.

The first example considered is an uniform cantilever AS4/3501-6 graphite-epoxy plate as shown in Figure 7.2 with material properties given in Table 5.2. Figures 7.8(a) and (b) show the snapshots of the axial velocities of the laminate with a symmetric ply orientation of $[0_8]$, at time instances $T = 250$ μs and $T = 375$ μs respectively. The plate dimensions are $L_X = 2.0$ m and $L_Y = 0.5$ m, and is modeled using single WSFE with $m = 64$ sampling points in Y direction. The impulse load as in Figure 5.2 is applied along edge AB in axial direction and the Y variation is obtained with $\alpha = 0.05$. The snapshot at $T = 250$ μs shows the forward moving axial wave, while at $T = 375$ μs the wave reflected from the fixed end of the cantilever plate can be captured. It should be mentioned here that the velocities at all the sampling points along Y direction and at any point along X direction used to obtain the snapshots are obtained from a single simulation. In Figures 7.9(a) and (b) the snapshots of the transverse velocities of the cantilever plate with a symmetric ply orientation of $[0_8]$ are presented for time instances $T = 500$ μs and $T = 1000$ μs respectively. The dimensions and the loading conditions are kept the same except that the loading here is applied in transverse direction. WSFE modeling is also performed with the similar parameters. Here, both the snapshots show the forward propagation of the flexural waves and the dispersive nature of these waves can also be observed.

The axial velocities at the mid point of edge AB of the cantilever graphite-epoxy plate used in the previous example with $L_X = 0.5$ m and $L_Y = 0.25$ m are plotted in Figure 7.10(a). The laminates have asymmetric ply lay-ups $[0_4/90_4]$ and $[0_4/60_4]$. These responses are due to the impulse load applied along AB in transverse direction and result from the axial-flexural coupling arising due to the asymmetric ply lay-up. The Y variation of the load is obtained with $\alpha = 0.03$. The amplitude of incident wave for $[0_4/90_4]$ is more than $[0_4/60_4]$ due to lower stiffness of the former ply lay-up. In Figure 7.10(b) the transverse velocities at mid-point of edge AB due to the load applied in axial direction are plotted. These responses result from the axial-flexural coupling. However, the amplitudes of these velocities arising due to the coupling vary considerably with the ply orientations and as expected are higher for $[0_4/90_4]$ due to its lower stiffness.

7.5.3   Wave propagation in ply-dropped plate

Ply dropping is done to taper a composite plate and is very much in use in many composite structures. This example of wave propagation in ply dropped plate due to impulse load is performed to emphasize the effectiveness of the present method in modeling such complex structures. The configuration of the plate is shown in Figure 7.11(a) and the overall dimension is $L_X = 1.5$ m and $L_Y = 0.5$ m. The plate is divided into three regions along X direction. From the fixed end to 0.5 m, the ply lay-up is $[0_4/90_4]$ and the total thickness is $h_1 = 0.01$ m. For the next 0.5 m, two plies are reduced. Here the thickness is $h_2 = 0.0075$ m and the lay-up is $[0_3/90_3]$. The last part has a thickness of $h_3 = 0.005$ m and the lay-up is $[0_2/90_2]$. As there are two discontinuities present, three WSFE are assembled to model the plate and the number of sampling points along Y direction is $m = 64$ with $\triangle Y = 0.008$ m. The unit impulse load is applied along edge AB and the Y variation of the load is obtained with $\alpha = 0.05$. In Figure 7.12(a), the axial velocity measured at the mid-point of edge AB due to load in axial direction is plotted. The axial velocity of corresponding uniform cantilever plate without ply drop and $L_X = 1.5$ m and $L_Y = 0.5$ m is also plotted for comparison. As expected, the amplitude of the incident wave for the uniform plate is considerably less than that for ply dropped plate. In addition, there are also much differences in the reflected waves arriving later. This is because, apart from the reflections from the lateral free edges and the fixed edge, the response of the ply dropped plate consists of reflections from the discontinuities present. These reflections are absent in the response of the uniform plate. In Figure 7.12(b), the corresponding transverse velocities due to transverse impulse load are plotted for both ply dropped and uniform plates. Similar to the axial velocities, here also there is a prominent difference in the amplitudes of the incident waves for the uniform and ply dropped plates. As explained earlier, the reflected waves also show considerable variations.

7.5.4   Wave propagation in folded plate structure

Next, wave propagation analysis in a folded plate structure, shown in Figure 7.11(c), is done using the formulated WSFE method. The folded plate model is obtained by assembling three two-dimensional WSFE. The number of sampling points along y direction is $m = 32$ for all the three plates. The plates have the same dimensions of $L_x = 0.5$ m and $L_y = 0.5$ m and a symmetric ply lay-up of $[0_8]$. The two slanted plates are at an angle of $60°$ to the X direction and the two ends denoted as AG and DH are considered fixed. Assembling is done along the local x direction (see Figure 7.11(c)) after the required transformation similar to conventional 1-D FE. The unit impulse load shown in Figure 5.2 is applied along edge BE and CF. The Y variation of the load is obtained with $\alpha = 0.1$. In Figures 7.13(a) and (b), the axial and transverse velocities measured at the mid-point of the edge BC, due to the above mentioned load applied in axial (X) and transverse (Z) directions, are plotted respectively. These numerical experiments show the capability of the present method to analyze the rather complicated folded plate structures.

7.6 Governing differential wave equations for axisymmetric cylinder

The governing differential equations for a hollow cylindrical structure [53] are generally written in cylindrical coordinate system $(r,\ \theta,\ z)$, which denotes the radial, the circumferential, and the axial directions respectively. The displacement components are u, v, and w in radial, circumferential, and axial directions respectively. For axisymmetric condition, the circumferential displacement v and the variations of the other two displacements with respect to θ are not considered in the modeling. Thus the equations of motion for an axisymmetric isotropic hollow cylinder contain only displacements u, w with variations along r, z (see Figure 7.14) and time t, and can be written as

$$(\lambda + 2\mu)\nabla_0^2\Delta = \rho\frac{\partial^2\Delta}{\partial t^2}$$

$$c_d^2\nabla_0^2\Delta = \frac{\partial^2\Delta}{\partial t^2}, \qquad c_d^2 = (\lambda + 2\mu)/\rho \tag{7.67}$$

$$\mu\nabla_1^2\mho_\theta = \rho\frac{\partial^2\mho_\theta}{\partial t^2}$$

$$c_s^2\nabla_1^2\mho_\theta = \frac{\partial^2\mho_\theta}{\partial t^2}, \qquad c_s^2 = \mu/\rho \tag{7.68}$$

where ∇_n^2 for subscripts $n = 0$ and 1 is defined as

$$\nabla_n^2 \equiv \frac{\partial^2}{\partial r^2} + \frac{1}{r}\frac{\partial}{\partial r} - \frac{n^2}{r^2} + \frac{\partial^2}{\partial z^2} \tag{7.69}$$

The variables Δ and $\mho_\theta$ are dilation and rotation vectors reduced for axisymmetric motions as

$$\Delta = \frac{1}{r}\frac{\partial(ru)}{\partial r} + \frac{\partial w}{\partial z} \tag{7.70}$$

$$\mho_\theta = \frac{\partial u}{\partial z} - \frac{\partial w}{\partial r} \tag{7.71}$$

and λ, μ are the Lame's constants and ρ is the mass density. From Equations 7.70 and 7.71, the displacements can be written in a decoupled form as

$$\nabla_1^2 u = \frac{\partial \Delta}{\partial r} + \frac{\partial \mho_\theta}{\partial z} \tag{7.72}$$

$$\nabla_0^2 w = \frac{\partial \Delta}{\partial z} - \frac{1}{r} \frac{\partial (r \mho_\theta)}{\partial r} \tag{7.73}$$

The associated boundary conditions are given as

$$\sigma_r = \lambda \Delta + 2\mu \frac{\partial u}{\partial r} \tag{7.74}$$

$$\sigma_z = \mu \frac{\partial u}{\partial z} + \frac{\partial w}{\partial r} \tag{7.75}$$

where σ_r and σ_z are the radial and axial forces respectively acting on the surface of the cylinder along z direction.

The governing PDEs and the associated boundary conditions derived here are reduced to a set of ODEs using Daubechies scaling function [20] approximation in time and one spatial dimension. The method has been explained in detail in the previous Section 7.3. The reduced equations derived from Equations 7.67 and 7.68 after temporal approximation are

$$c_d^2 \left[\frac{\partial^2}{\partial r^2} + \frac{1}{r} \frac{\partial}{\partial r} + \frac{\partial^2}{\partial z^2} \right] \widehat{\Delta}_j = -\gamma_j^2 \widehat{\Delta}_j \tag{7.76}$$

$$c_s^2 \left[\frac{\partial^2}{\partial r^2} + \frac{1}{r} \frac{\partial}{\partial r} - \frac{1}{r^2} + \frac{\partial^2}{\partial z^2} \right] \widehat{\mho}_{\theta j} = -\gamma_j^2 \widehat{\mho}_{\theta j} \quad j = 0,\, 1,\, \ldots, n-1 \tag{7.77}$$

Similarly, the transformed forms of the decoupled displacement equations (Equations 7.72 and 7.73) and force boundary conditions (Equations 7.74 and 7.75) can be written as

$$\nabla_1^2 \widehat{u}_j = \frac{\partial \widehat{\Delta}_j}{\partial r} + \frac{\partial \widehat{\mho}_{\theta j}}{\partial z} \tag{7.78}$$

$$\nabla_0^2 \widehat{w}_j = \frac{\partial \widehat{\Delta}_j}{\partial z} - \frac{1}{r} \frac{\partial (r \widehat{\mho}_{\theta j})}{\partial r} \tag{7.79}$$

$$\widehat{\sigma}_{rj} = \lambda \widehat{\Delta}_j + 2\mu \frac{\partial \widehat{u}_j}{\partial r} \tag{7.80}$$

$$\widehat{\sigma}_{zj} = \mu \frac{\partial \widehat{u}_j}{\partial z} + \frac{\partial \widehat{w}_j}{\partial r} \tag{7.81}$$

Following a scaling function transformation in z and decoupling through eigenvalue analysis considering un-restrain boundary conditions, the Equations 7.76 and 7.77 can be further reduced to ODEs as

$$c_d^2 \left[\frac{d^2}{dr^2} + \frac{1}{r} \frac{d}{dr} - \beta_i^2 \right] \widetilde{\Delta}_{ij} = -\gamma_j^2 \widetilde{\Delta}_{ij} \quad i = 0,\, 1,\, \ldots,\, m-1 \tag{7.82}$$

$$c_s^2 \left[\frac{d^2}{dr^2} + \frac{1}{r}\frac{d}{dr} - \frac{1}{r^2} - \beta_i^2 \right] \widetilde{\mho}_{\theta ij} = -\gamma_j^2 \widetilde{\mho}_{\theta ij} \qquad (7.83)$$

Similarly the decoupled displacement equations given by Equations 7.78, 7.79, and the boundary conditions Equations 7.80, 7.81 can be written as

$$\left[\frac{d^2}{dr^2} + \frac{1}{r}\frac{d}{dr} - \frac{1}{r^2} - \beta_i^2 \right] \widetilde{u}_{ij} = \frac{d\widetilde{\Delta}_{ij}}{dr} - \imath\beta\widetilde{\mho}_{\theta ij} \qquad (7.84)$$

$$\left[\frac{d^2}{dr^2} + \frac{1}{r}\frac{d}{dr} - \beta_i^2 \right] \widetilde{w}_{ij} = -\imath\beta_i\widetilde{\Delta}_{ij} - \frac{1}{r}\frac{d(r\widetilde{\mho}_{\theta ij})}{dr} \qquad (7.85)$$

$$\widetilde{\sigma}_{rij} = \lambda\widetilde{\Delta}_{ij} + 2\mu\frac{d\widetilde{u}_{ij}}{dr} \qquad (7.86)$$

$$\widetilde{\sigma}_{zij} = -\imath\beta_i\mu\widetilde{u}_{ij} + \frac{d\widetilde{w}_{ij}}{dr} \qquad (7.87)$$

The subscript i and j are dropped hereafter for simplified notations. All the equations derived however have to be evaluated for $i = 0, 1, \ldots, n-1$ and $j = 0, 1, \ldots, m-1$.

7.7 Bessel function solution for axisymmetric cylinder

The degrees of freedom associated with element formulation are shown in Figure 7.14. The element has two degrees of freedom per node, which are $\widetilde{u}$ and $\widetilde{w}$. The two sets of decoupled ODEs given by Equations 7.82 and 7.83 and the displacement relations Equations 7.84 and 7.85 are to be solved for $\widetilde{u}$ and $\widetilde{w}$. The actual solutions $u(r, z, t)$ and $w(r, z, t)$ are obtained using inverse wavelet transform twice for temporal and spatial (axial) dimensions [100].

Here, the Bessel's functions are used for solution of the transformed governing equations given by Equations 7.82 and 7.83. The solutions for $\widetilde{\Delta}$ and $\widetilde{\mho}_\theta$ are obtained as

$$\widetilde{\Delta}(r) = C_1 J_0(k_e r) + C_2 Y_0(k_e r) \qquad (7.88)$$

$$\widetilde{\mho}_\theta(r) = C_3 J_1(k_s r) + C_4 Y_1(k_s r) \qquad (7.89)$$

where $k_e^2 = (\gamma^2/c_d^2 - \beta^2)$, $k_s^2 = (\gamma^2/c_s^2 - \beta^2)$, and C_1, C_2, C_3, C_4 are constants. J_n and Y_n are Bessel's functions of first and second kinds.

Substituting Equations 7.88, 7.89 in Equations 7.84, 7.85 and solving for $\widetilde{u}$ and $\widetilde{u}$ gives [39]

$$\widetilde{u}(r) = -k_e[C_1 J_1(k_e r) + C_2 Y_1(k_e r)] - \imath\beta[C_3 J_1(k_s r) + C_4 Y_1(k_s r)] \quad (7.90)$$

$$\widetilde{w}(r) = -\imath\beta[C_1 J_0(k_e r) + C_2 Y_0(k_e r)] - k_s[C_3 J_0(k_s r) + C_4 Y_0(k_s r)] \quad (7.91)$$

The above solutions provide the interpolating functions for forming the elemental dynamic stiffness matrix. The unknown constants $\{a\} = \{C_1, C_2, C_3, C_4\}$ can be determined from the transformed nodal displacements $\widetilde{u}(r)$, $\widetilde{w}(r)$ at inner $(r = r_i)$ and outer $(r = r_o)$ radii. The nodal displacement vector is $\{\widetilde{\mathbf{u}}^{\mathbf{e}}\} = \{\widetilde{u}_1, \widetilde{w}_1, \widetilde{u}_2, \widetilde{w}_2\}$, where $\widetilde{u}_1 = \widetilde{u}(r_i)$, $\widetilde{w}_1 = \widetilde{w}(r_i)$, $\widetilde{u}_2 = \widetilde{u}(r_o)$ and $\widetilde{w}_2 = \widetilde{w}(r_o)$. Thus we can relate nodal displacements and unknown constants from Equations 7.90 and 7.91 as

$$\{\widetilde{\mathbf{u}}^{\mathbf{e}}\} = [\mathbf{T_1}]\{a\} \tag{7.92}$$

Substituting Equations 7.90 and 7.91 in the transformed force boundary conditions given by Equations 7.86 and 7.87, they can be written as

$$\begin{aligned}
\widetilde{\sigma}_r(r) &= [-\{(\lambda + 2\mu)k_e^2 - \lambda\beta^2\}J_0(k_e r) + 2\mu k_e^2 J_1(k_e r)/(k_e r)]C_1 \\
&+ [-\{(\lambda + 2\mu)k_e^2 - \lambda\beta^2\}Y_0(k_e r) + 2\mu k_e^2 Y_1(k_e r)/(k_e r)]C_2 \\
&+ 2\imath\mu k_s\beta[J_0(k_s r) - J_1(k_s r)/(k_s r)]C_3 \\
&+ 2\imath\mu k_s\beta[Y_0(k_s r) - Y_1(k_s r)/(k_s r)]C_4 \tag{7.93} \\
\widetilde{\sigma}_z(r) &= -2\imath\mu\beta k_e[J_1(k_e r)C_1 + Y_1(k_e r)] \\
&- \mu(k_s^2 - \beta^2)[J_1(k_s r)C_3 + Y_1(k_s r)C_4] \tag{7.94}
\end{aligned}$$

From the two above equations, the nodal force vector $\{\mathbf{F^e}\} = \{\widetilde{\sigma}_{r1}, \widetilde{\sigma}_{z1}, \widetilde{\sigma}_{r2}, \widetilde{\sigma}_{z2}\}$ can be related to the constants $\{a\}$ as

$$\{\widetilde{\mathbf{F}}^{\mathbf{e}}\} = [\mathbf{T_2}]\{a\} \tag{7.95}$$

where $\widetilde{\sigma}_{r1} = \widetilde{\sigma}_r(r_i)$, $\widetilde{\sigma}_{z1} = \widetilde{\sigma}_z(r_i)$, $\widetilde{\sigma}_{r2} = \widetilde{\sigma}_r(r_o)$ and $\widetilde{\sigma}_{z2} = \widetilde{\sigma}_z(r_o)$. Finally from Equations 7.92 and 7.95, a relation between transformed nodal forces and displacements is obtained as

$$\{\widetilde{\mathbf{F}}^{\mathbf{e}}\} = [\mathbf{T_2}][\mathbf{T_1}]^{-1}\{\widetilde{\mathbf{u}}^{\mathbf{e}}\} = [\widetilde{\mathbf{K}}^{\mathbf{e}}]\{\widetilde{\mathbf{u}}^{\mathbf{e}}\} \tag{7.96}$$

where $[\widetilde{\mathbf{K}}^{\mathbf{e}}]$ is the exact elemental dynamic stiffness matrix.

7.8   Wave propagation in isotropic axisymmetric cylinders

In this section, the developed two-dimensional WSFE is used to analyze axisymmetric radial and axial wave propagations in isotropic cylinders due to the broad-band impulse excitation. First, the simulations are performed for an aluminum hollow cylinder free at both ends and internal pressures acting in radial and axial directions. The cylinder has an inner radius $r = r_i$, outer radius $r = r_o$, and the axial length is L_z, as shown in Figure 7.15(a).

The loading condition is shown in Figures 7.16(a) and (b) for radial and axial pressures respectively. Next, a more complex structure of a bi-material, i.e., aluminum-steel (see Figure 7.15(b)) cylinder is considered for analysis. The material properties of the aluminum and steel cylinders are considered respectively as follows, Young's modulus $E_a = 70$ GPa and $E_s = 200$ GPa, mass densities $\rho_a = 2700$ kg/m^3 and $\rho_s = 7860$ kg/m^3, and Poisson's ratio $\nu = 0.3$. Here, the spatial sampling rate $\triangle z$ is varied depending on L_Z. As mentioned earlier, for the aluminum cylinder in Figure 7.15(a), only one 2-D WSFE is used to simulate the responses, independent of its dimensions. For bi-material cylinder in Figure 7.15(b), two elements are required for modeling due to the discontinuity present in the problem.

7.8.1   Wavenumber computation

The spectrum relations for an aluminum hollow cylinder with $r_i = 0.02$ m and $r_o = 0.03$ m at different radii r obtained from the wavenumber computation discussed earlier are plotted in Figures 7.17(a) to (c). Figures 7.17(a) to (c), respectively, show the real or propagating part of the wavenumbers for $r = 0.02$, 0.025, and 0.03 m. These wavenumbers have significant imaginary parts which imply that the waves are inhomogeneous in nature; in other words, they attenuate while they propagate. As the radius is moved from inner to outer radius, the cut-off frequency which is the frequency at which the wavenumber is zero is also driven outward to a higher frequency. The wavenumbers are plotted for an axial (Z) wavenumber of $\beta = 50$ and are obtained with $\triangle t = 4$ μs, i.e., for Nyquist frequency of $f_{nyq} = 125$ kHz.

7.8.2   Time domain analysis

Here, first the responses obtained using the formulated element are validated with two-dimensional FE analysis. In Figures 7.18(a) and (b), the radial velocities for an aluminum cylinder with $r_i = 0.05$ m, $r_o = 0.1$ m, and $L_z = 2.0$ m are plotted and compared with FE results. Figures 7.18(a) and (b) show the radial velocities at mid-point (at $L_z/2$ from free ends) and at inner $r = r_i$ and outer radii $r = r_o$ respectively. Load applied is the unit impulse as discussed earlier and shown in Figure 5.2. It is applied as an uniform pressure P_i (see Figure 7.16(a)) in the radial direction along the inner surface ($r = r_i$). As an example, such internal pressure can be encountered in fluid carrying pipes and other cylindrical pressure vessels. The FE results are obtained using ANSYS 10.0, with 10220, 4-noded quadrilateral axisymmetric element (PLANE42) mesh. Newmark's scheme with time step 2 μs is used for time integration. A further refinement of the FE mesh does not give significant difference in results. WSFE for this example is formulated with $N = 22$, time interval $\triangle t = 2$ μs, and time window $T_w = 512$ μs. Similar axial velocity responses are plotted in Figures 7.19(a) and (b) under same loading conditions as before except that here they are applied in axial direction (see Figure 7.16(b)). Figures 7.19(a) and (b) show the axial velocities at $r = r_i$ and $r = r_o$ at distances $L_z/2$ from the free ends respectively. Even here, the responses are validated with FE results and it can be seen that the responses compare very well. The modeling parameters for WSFE and the FE mesh are kept the same as in the previous case of radial velocities. Next, WSFE is used to model a relatively complex structure of bi-material, i.e., aluminum-steel cylinder. The configuration is shown in Figure 7.15(b), where the inner cylinder is composed of aluminum and is surrounded by a concentric steel cylinder. The innermost radius is $r_i = 0.05$ m, radius of the interface is $r_m = 0.09$ m, and the outermost radius is $r_o = 0.1$ m. Length of the cylinder is $L_z = 2.0$ m and the impulse load (see Figure 5.2) is applied uniformly at the inner surface $r = r_i$ as an internal pressure in radial/axial directions. As stated earlier, two WSFE are required to model the structure because of the presence of discontinuity. WSFE modeling involves $m = 64$ spatial sampling points. In Figures 7.20(a) and (b), the radial and axial velocities measured at the mid-point $(L_z/2)$ and $r = r_i$ are plotted. The responses are also compared with those of an aluminum cylinder of same dimensions. The amplitude of the incident wave for the aluminum cylinder is much more than that of the aluminum-steel cylinder as expected as the latter has a much higher stiffness. Comparisons are also provided with FE results for the aluminum-steel cylinder. The FE mesh and the Newmark's time integration scheme are the same as the previous example. It can be seen that the responses match well with the FE results for both radial and axial wave propagations.

In brief, the formulation of two-dimensional WSFE for finite dimensional waveguides are presented in this chapter. The method use Daubechies scaling function transform in temporal and one spatial dimension. The method is then used to simulate time and frequency domain wave propagation in isotropic plate under in-plane and out-of-plane loading, and axisymmetric cylinders. Problems of higher complexities like plate with ply-drops, folded plate structure, and anisotropic composite plates are modeled to emphasize the efficiency of this wavelet-based technique. In the next two chapters, the one- and two-dimensional WSFE presented so far will be implemented to model single- and multi-walled carbon nanotubes and nano-composites as continuum beams and shells.

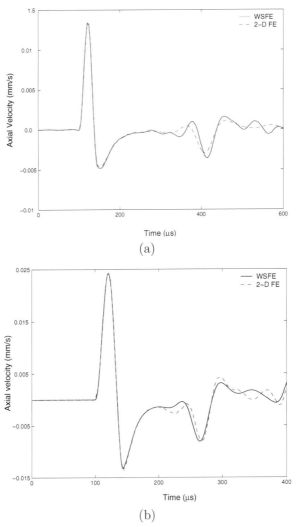

FIGURE 7.6: Axial velocity of fixed-fixed cantilever plate (see Figure 7.2(a)) with $L_X = 4.0$ m, (a) $L_Y = 0.5$ m and (b) $L_Y = 0.25$ m measured at mid-point of free end AB

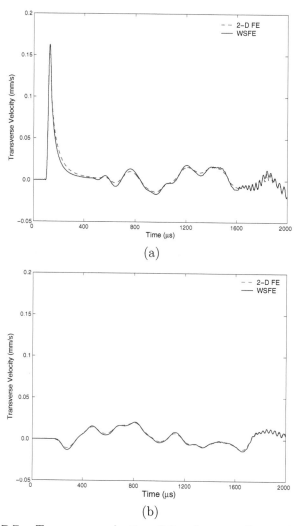

FIGURE 7.7: Transverse velocity of free-free cantilever plate (see Figure 7.2(a)) with $L_X = 4.0$ m and $L_Y = 0.5$ m as measured at (a) mid and (b) quarter points of the free end AB

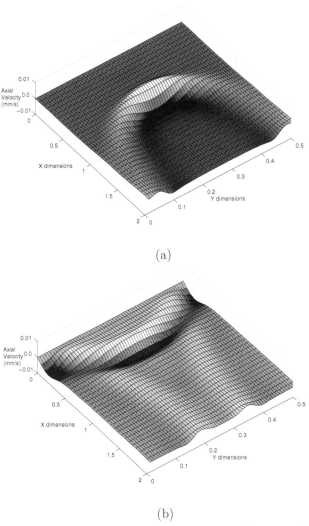

(a)

(b)

FIGURE 7.8:   Snapshots of axial velocities at time instances (a) $T = 250\ \mu$s and (b) $T = 375\ \mu$s in a $[0]_8$ cantilever plate due to tip impulse load applied in axial direction

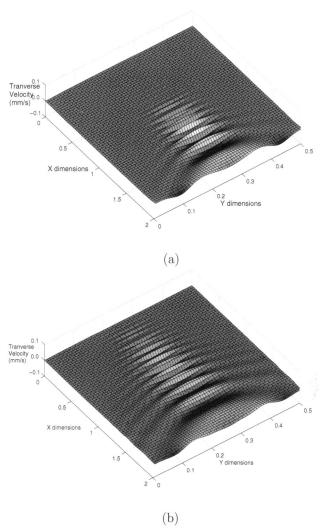

(a)

(b)

FIGURE 7.9:   Snapshots of transverse velocities at time instances (a) $T = 500$ μs and (b) $T = 1000$ μs in a $[0]_8$ cantilever plate due to tip impulse load applied in transverse direction

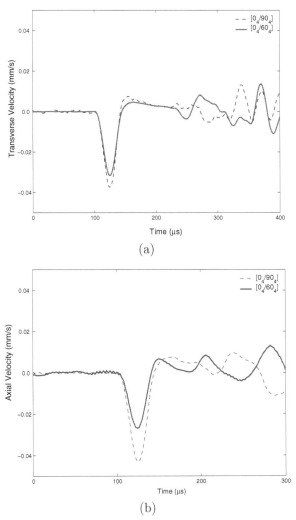

FIGURE 7.10:   (a) Axial and (b) transverse velocities at mid-point of edge AB in asymmetric graphite-epoxy laminates due to tip impulse load applied in transverse and axial directions along AB, respectively

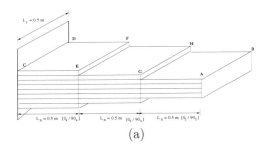

(a)

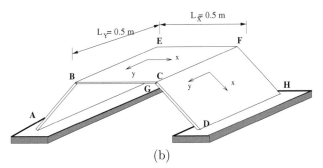

(b)

FIGURE 7.11:   (a) Ply drop and (b) folded plates

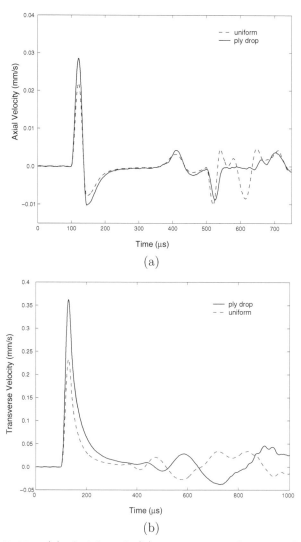

FIGURE 7.12: (a) Axial and (b) transverse velocities of asymmetric $[0_4/90_4]$ uniform and ply dropped plate

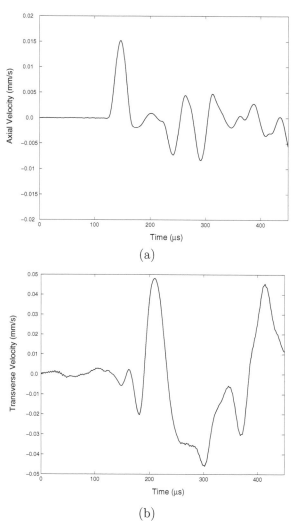

FIGURE 7.13: (a) Axial and (b) transverse velocities at mid-point of BC of $[0_8]$ folded plate (Figure 7.11(c)) due load applied along edge BE and CF in axial and transverse directions, respectively

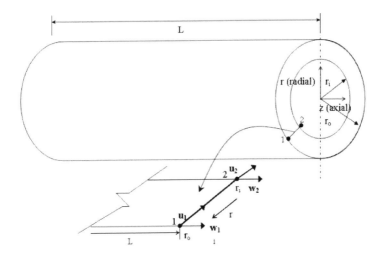

FIGURE 7.14: Axisymmetric element with nodal displacements and forces

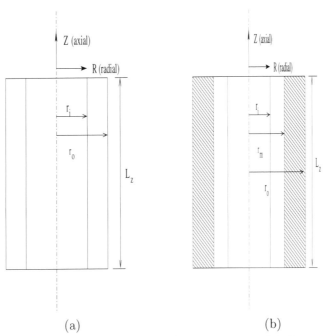

FIGURE 7.15: (a) Aluminum and (b) aluminum-steel hollow cylinders

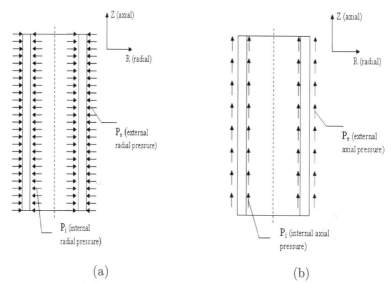

(a)   (b)

FIGURE 7.16:   (a) Radial and (b) axial internal and external pressure

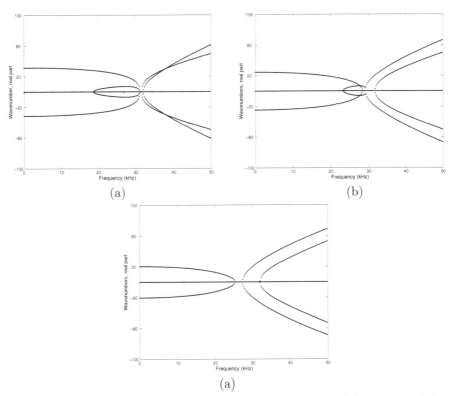

FIGURE 7.17: Real parts of wavenumbers at (a) $r = r_i$, (b) $r = (r_i + r_o)/2$, and (c) $r = r_o$

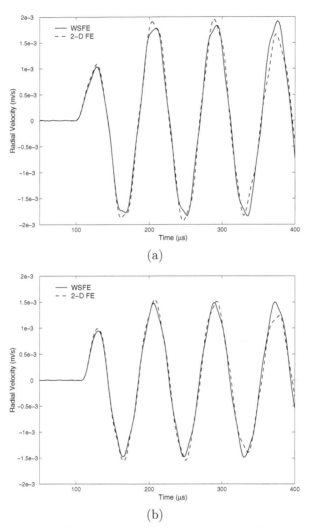

FIGURE 7.18: Radial velocities at mid-point along the axial direction at (a) $r = r_i$ and (b) $r = r_o$ due to internal radial pressure

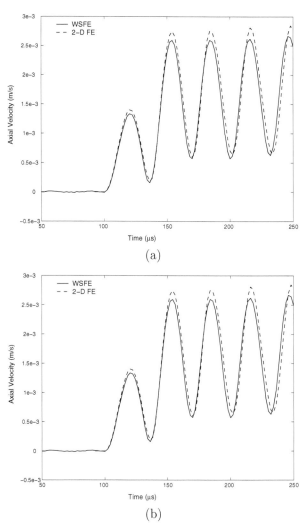

FIGURE 7.19: Axial velocities at mid-point along the axial direction at (a) $r = r_i$ and (b) $r = r_o$ due to internal axial pressure

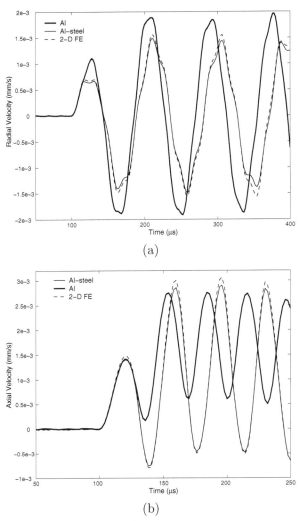

FIGURE 7.20:   (a) Radial and (b) axial velocities in bi-material Al-steel cylinder at mid-point along axial direction at $r = r_i$

Chapter 8

Vibration and Wave Propagation in Carbon Nanotubes

This chapter explains the application of WSFE developed for one- and two-dimensional nanotube waveguides to study vibration and wave properties of carbon nanotubes (CNTs). Over the last decade, with the advent of nano-structured material, a large group of researchers in computational mechanics are directing their efforts to understand the mechanical properties of such structures. This is primarily because these structures possess immense potential to serve as excellent structural material with enhanced functionality. Here, WSFE is used to model CNTs using continuum shell and beam theories to obtain vibrational and wave characteristics. The models are developed for both single walled and multi-walled carbon nanotubes.

8.1 Carbon nanotubes: introduction

After their discovery in 1991 by Ijima [41], a great deal of theoretical and experimental studies [43] have been directed toward understanding the properties of carbon nanotubes (CNTs) due to their enormous applications. These applications have been foreseen because of the superior physical properties which include large stiffness (of the order of 1 TPa), improved thermal (stable up to 2800° C) and electrical (1000 times of copper) conductivity, and low density (1/6 of steel). These properties result from the symmetric structures of CNTs which can be thought of as graphene sheets rolled into a tube. Graphene is formed as 2-D sheet of carbon atoms arranged in a hexagonal array. The atomic structure of the nanotube is characterized by the chiral vector $\vec{C}_h$ and chiral angle θ. The size, mechanical, thermal, and electrical properties of CNTs are dependent on these atomic structures. The structures of CNTs are mainly of two types, zig-zag and armchair. The armchair nanotubes exhibit better ductility and conductivity than zig-zag CNTs. Apart from single-walled carbon nanotubes (SWNTs) having hollow cylindrical configuration, CNTs are also formed with the structures of several coaxial SWNTs held together with weak van der Waals forces and these are called multi-walled carbon nanotubes (MWNTs). The dimensions of CNTs are within the following ranges, thickness 0.0066 to 0.34 nm, radius 0.4 to 100 nm, and length varying between 1 nm to 1 μm. The distance between the walls of MWNTs is in general approximately 0.34 nm. As said before, the high strength of CNTs is a powerful motivation for their use as reinforcing materials to form high strength, lightweight composites generally referred to as nano-composites. Embedding of CNTs will help to enhance the damping characteristics, increase buckling capacity, and improve resistance to higher temperature (re-entry vehicle).

Parallel to the extensive experimental study to understand the mechanical behavior of CNTs, equal effort is being given to develop computational models. These theoretical studies can be broadly divided into atomistic modeling and continuum modeling. The most popular atomistic modeling techniques include molecular dynamics (MD), tight binding molecular dynamics (TBMD) and density functional theory (DFT). Although these atomistic simulations have been used successfully to simulate the properties of CNTs, the associated large computational cost restricts their use particularly for large scale simulations. This limitation of atomistic modeling encouraged researchers to look into the suitability of using continuum mechanics to model CNTs and nano-composites. Such continuum modeling has been proved to be efficient in predicting several mechanical properties of these nano-structures. Different beam theories, linear and non-linear shell theories, and non-local elasticity approaches have been adopted for modeling CNTs considering them to be continuum [90, 95, 98, 99, 104].

The vibration and wave propagation analysis of CNTs [35] are relevant due their various applications [30] which include sensors [51, 50], actuators [32, 5], resonator and oscillator [49, 87, 107], transport, and optical phenomena. Even for such analysis, use of both atomistic and continuum models are presented in literature. It is well known that CNTs can propagate waves of very high frequency content (of the order of Tera-Hertz). At such high frequencies, continuum models based on FE methods cannot be adopted due to their limitation of limiting the element size to that of wavelength, which is very small at such frequencies. Hence, spectral finite element method is an ideal candidate for such situations. In addition, SFE allows to study the frequency domain wave characteristics which is not possible with other techniques. This study for CNTs will bring out several wave propagation features, which are not observed in cases of macro-scale structures.

8.2 Axisymmetric shell model of single-walled carbon nanotubes

In this chapter, first the modeling of single-walled carbon nanotube (SWNT) is done by assuming it as an axisymmetric cylinder [66]. The two-dimensional WSFE model for such structures has already been developed in the last chapter, for wave propagation analysis in aluminum cylinders. Similar element is used here, except that it is used at nano-scale to study high frequency and wave propagation characteristics in SWNT. Thus the formulation of two-dimensional WSFE is not repeated in this chapter and only numerical experiments are presented and analyzed.

8.2.1 Frequency domain analysis

Here numerical experiments are performed to study the effects of thickness, $2h$ and radius, R of the SWNT on the higher axisymmetric vibrational modes. The vibration includes predominantly radial breathing modes (RBMs) and longitudinal modes, and coupled radial-longitudinal modes. Here the uncoupled pure twisting mode resulting from circumferential displacement is not considered. The bulk material properties are Young's modulus $Eh = 360 \text{ J/m}^2$ [102], mass density $\rho h = 2270 \times 0.34 \text{ kg/m}^3$, and Poisson's ratio $\nu = 0.2$ [99]. In particular these parameters are not dependent on the definition of the thickness $2h$ [102]. The variation of Young's modulus with the radius is not well defined. In reference [54] it has been stated that the Young's modulus does not vary with the radius, while a variation has been predicted in references [78, 12]. Considering uniform Young's modulus irrespective of radius varying between 0.3-1.0 nm, the fundamental RBM frequencies are found to follow the 1/2R law in tune with the experimental results [15]. Thus in this work, the Young's modulus is considered to be independent of radius. The radius R of (N, N) SWNT is calculated as $R = 3Na_{C-C}/(2\pi)$ [14], where a_{C-C} is carbon bond length and equal to 0.142 nm. The inner and outer radii are referred to as $a = R - h$ and $b = R + h$ respectively. The frequencies are expressed in optical units cm^{-1}, where the conversion is given as 1 Hz $= 3.336 \times 10^{-11}$ cm^{-1}.

First validation of the developed SWNT model is done by comparing the simulated results with those available in literature. In Figure 8.1, the acoustic phonon dispersion relation of a $(10, 10)$ SWNT with $R = 0.678$ nm and $2h = 0.09$ nm is presented for the axisymmetric condition. Comparisons are made with the corresponding results ($n = 0$) obtained from *abinitio* [103] and 3-D elasto-dynamic continuum [15] models. In the figure, the frequencies in optical unit are presented for varying longitudinal wavenumber K_z normalized as $K_z a$. In the present formulation K_z represents β as in Equation 7.82. Similar to reference [55, 15], the dispersion relation is obtained by equating

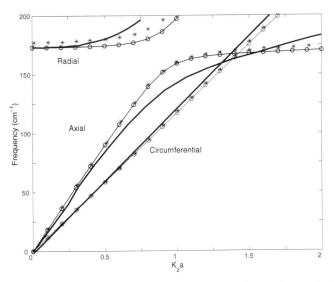

FIGURE 8.1: Axisymmetric acoustic phonon dispersion relation for a $(10, 10)$ SWNT with $R = 0.678$ nm and $2h = 0.09$ nm ('$-$o$-$' present model, '$*$' continuum model Ref. [15] and '$-$' abinitio model Ref. [103])

the determinant of the matrix $[C]$ in Equation 7.95 to zero and solving for γ, for different values of β, i.e., K_z. Though the radial and longitudinal displacements are only considered in this work, here the axisymmetric circumferential displacement is also studied and compared. It can be seen that the acoustic phonon dispersion relation predicted by the present model is in good agreement with that obtained from references [103, 15].

Next, the higher axisymmetric vibrational modes are studied to understand the effect of geometrical parameters on these higher natural frequencies. Figure 8.2(a) shows the variation of the first two RBMs with $2h$ for a SWNT of $R = 0.678$ nm and length $L = 25$ nm. The wall thickness is varied from $2h = 0.24$ nm to 0.34 nm. It can be seen that the first frequency of RBM is not dependent on the SWNT wall thickness considered in the continuum model. This is in tune with that stated in reference [15] and other theoretical work [55, 98]. The first RBM frequency calculated is 179 cm^{-1} which compares well with the value 182 cm^{-1} derived in reference [15] and is close to the experimental value [81], 186 cm^{-1}. A better match between the results can be obtained by fitting the bulk material properties like Young's modulus, mass density, and Poisson's ratio. However, interestingly, it has been observed that the higher RBM frequencies vary considerably with the SWNT wall thickness. One explanation for this behavior is that at higher RBM modes, the deformation vary along the thickness of the SWNT. In other words, at higher frequencies, local deformation of the SWNT wall occurs along with the overall radial breathing deformation. This makes the higher RBM frequencies dependent on the wall thickness. Apart from this, at higher frequencies the effect of radial-longitudinal coupling is more prominent which may also be a reason for such behavior. However, this observation is of much importance as it suggests that in continuum modeling the wall thickness has to be fitted from the experimental results to correctly predict the higher RBM frequencies unlike the first RBM frequency. From Figure 8.2(a), it can be seen that the second RBM frequencies follow a linear variation with wall thickness and have nearly similar slopes.

In Figure 8.2(b), the first two RBM frequencies of SWNT are plotted for varying radius R from 0.3 nm to 1.0 nm. The length of SWNT is $L = 25$ nm and $2h = 0.34$ nm. It can be seen that the variation of first RBM frequency is linear and obeys the $1/2R$ rule. This has been predicted in reference [42]. Here, the variation of the first RBM frequency is obtained as 243 cm^{-1} (nm/2R) which compares well with the value 246 cm^{-1} (nm/2R) obtained in reference [15]. Even here, the higher RBM frequencies show a counterintuitive behavior. These frequencies do not vary significantly with the increase in the R as seen from the figure. Again the complicated nature of the higher frequencies due to the local deformation of SWNT wall and increased effect of longitudinal-radial mode coupling may be an explanation of this behavior.

However, it should be mentioned here that the characteristics of higher RBM frequencies have been studied within a certain region of $2h$ and R. Though in this work the viable ranges of $2h$ and R have been considered, it is probable that the trend may vary over other ranges of these parameters. Furthermore, it has been observed by performing numerical experiments that for a given R, the length L of SWNT has no significant effect on these frequencies except for very low slenderness (L/R) ratio which is not common in carbon nanotubes.

Figure 8.3(a) show the frequencies of the first two longitudinal vibrational modes of a SWNT with $L = 25$ nm, $R = 0.678$ nm, and $2h$ varied over 0.24 to 0.34 nm. It can be seen that both the frequencies have a linear variation with the wall thickness. Such variation is much justified as the longitudinal stiffness and thus the corresponding frequency is primarily dependent on the wall thickness. The slopes of the plots are nearly equal for both the modes. The rate of decrease of the longitudinal frequencies with the increase of $2h$ is high and hence only a small range is considered for the study. In addition, the frequencies of the longitudinal modes are much higher than the corresponding radial modes. As an example for (10,10) SWNT with $R = 0.678$ nm and $2h = 0.34$ nm, the frequency of fundamental longitudinal mode is 692 cm^{-1} while that of fundamental RBM is 179 cm^{-1}. Thus for determining the longitudinal vibrational frequencies using continuum model, the wall thickness plays an important role even for the first mode. Hence, this parameter should also be fitted with the experimental results. In Figure 8.3(b), the first two longitudinal mode frequencies of the SWNT with $L = 25$ nm and $2h = 0.34$ nm are plotted for different R. It is observed that these frequencies are not

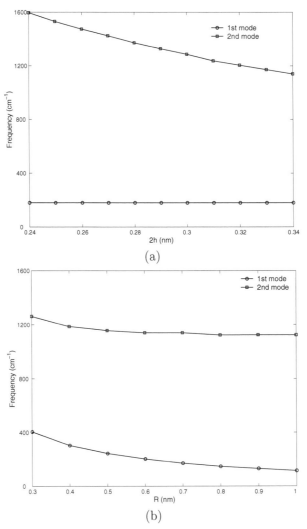

FIGURE 8.2:   Variation of frequencies of SWNT with (a) wall thickness $2h$ and (b) radius R, for first five radial vibrational modes

influenced by the R. Similar to the RBMs, even here the length has no effect on the frequencies of longitudinal mode for $L/R < 10$.

The frequencies of radial and longitudinal modes presented in the last two examples though have the effect of the radial-longitudinal coupling particularly at the higher mode; they are predominantly radial or longitudinal modes. However, there exists a mode which results purely from the coupling as has also been reported in literature [15, 14] and this causes the coupled vibration, i.e., a radial vibration occurs due to purely axial excitation and vice versa. In Figure 8.4(a), the dependence of the first two frequencies of this coupled mode with thickness $2h$ is shown for a SWNT with $R = 0.678$ nm and $L = 25$ nm. The wall thickness are varied from $2h = 0.24$ to 0.34 nm. Even here, the first frequency is invariant to $2h$, while the other two higher frequencies have linear variation. The frequency of the first coupled mode is very near to the first RBM frequency while the other higher modal frequencies vary considerably. Figure 8.4(b) shows the frequencies of radial-longitudinal coupled mode of a SWNT with $2h = 0.34$ nm and $L = 25$ nm for varying R from 0.3 to 1.0 nm. Again here, the trend of variation is similar to that of the RBM frequencies. This coupled vibrational mode is also independent of L for a slenderness ratio $L/R < 10$.

8.2.1.1 Time domain analysis

The time domain analysis of responses of carbon nanotubes to excitation signals has wide scope and applications in sensor development. Unlike the Fourier transform-based analysis, the present wavelet-based formulation allows accurate simulations of time domain responses along with the frequency domain analysis provided earlier. In this section, the effect of wall thickness and radius of SWNT on the responses are studied. The SWNT considered has a length of $L = 25$ nm and is free-free at ends. A very short duration (0.05 ps) impulse load with frequency content $0 - 45$ THz is applied at mid-point of the SWNT along the outer surface, $r = b$, in radial or axial direction. As a result of the radial-longitudinal coupling present, a loading in radial direction results in longitudinal deformation and correspondingly the axial loading causes radial deformation. In the present case, the excitation is high frequency broadband and hence apart from the first mode, the higher modes participate in the response. Thus for such excitation, the continuum model should be able to capture the higher vibrational modes of the SWNT accurately. In addition, due to the participation of the higher frequencies, even when the parameters like wall thickness and radius do not effect the frequencies of a single vibrational mode largely, they may influence the overall time domain response to a great extent. This has been emphasized through numerical experiments, which show that the geometrical parameters of SWNT have a prominent influence on the time domain response.

In Figures 8.5(a) and (b), the snapshots of radial wave propagation in SWNT with different R and $2h$ are shown at time instances $T_1 = 0.5$ ps and

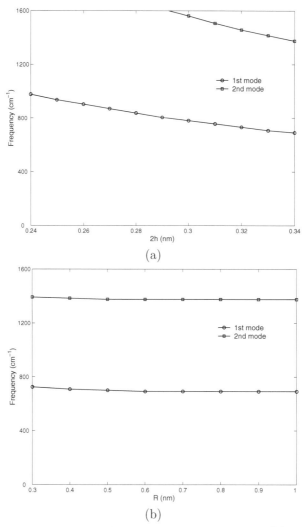

FIGURE 8.3:   Variation of frequencies of SWNT with (a) wall thickness $2h$ and (b) radius R, for first five longitudinal vibrational modes

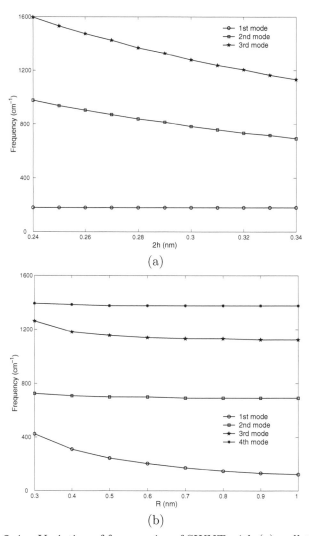

FIGURE 8.4: Variation of frequencies of SWNT with (a) wall thickness $2h$ and (b) radius R, for first five longitudinal radial coupled vibrational modes

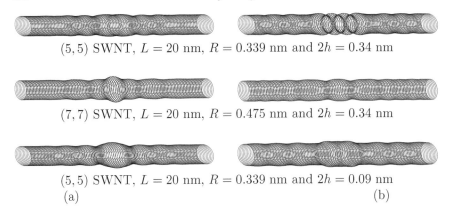

(5, 5) SWNT, $L = 20$ nm, $R = 0.339$ nm and $2h = 0.34$ nm

(7, 7) SWNT, $L = 20$ nm, $R = 0.475$ nm and $2h = 0.34$ nm

(5, 5) SWNT, $L = 20$ nm, $R = 0.339$ nm and $2h = 0.09$ nm

(a)   (b)

FIGURE 8.5:   Radial wave propagation snapshots at instances (a) 0.5 ps and (b) 1.0 ps

$T_2 = 1.0$ ps respectively. The snapshots are presented for the deformations at $r = R$. These wave propagations result from the impulse loading as described earlier and applied at mid-point of SWNT in radial direction at $T_0 = 0.1$ ps. The results are presented for three SWNT configurations with $R = 0.339$ and $2h = 0.34$ nm, $R = 0.475$ and $2h = 0.34$ nm, and $R = 0.339$ and $2h = 0.09$ nm. It can be observed that changing the wall thickness from $2h = 0.34$ nm to $2h = 0.09$ nm for a (5, 5) SWNT, the vibration pattern changes considerably, though it has been observed from the study of RBM frequencies earlier (see Figure 8.2(a)) that the frequencies of only higher RBMs are affected by the parameters $2h$. This may be again justified by the explanation given in the last paragraph. Similar difference in the wave transmission snapshots is observed when the radius of SWNT is increased from $R = 0.339$ nm to 0.475 nm. This difference may have resulted from the effect of R on the first RBM frequency (see Figure 8.2(b)).

8.3   Thin shell model of multi-walled carbon nanotubes

Apart from axisymmetric and three-dimensional shell models of CNTs, substantial work on simplified Flügge's thin shell model of CNTs has been reported in literature. While in the axisymmetric models, displacements can be considered to vary of axial and radial dimensions, in thin shell theories, the displacements vary only in axial and circumferential directions. Thus, the axisymmetric and three-dimensional elasto-dynamic shell model can be used for any arbitrary wall thickness of CNTs while Flügge's shell theory [99] is based on thin wall assumption. Flügge's shell theory however has been widely used for buckling, vibration, and wave propagation analysis of SWNT, MWNT, and also CNT embedded nano-composites.

In this section, MWNTs are modeled as continuum cylindrical shells using Flügge's theory, with the walls coupled through inter-wall van der Waals (vdW) forces. The model is developed considering variation only along the circumferential (θ) direction, while displacements along axial, radial, and circumferential directions are considered. The WSFE model is then used to study wave properties of single, double, and three walled carbon nanotubes. The properties include wavenumbers, wave speeds, and time history of the wave response due to high frequency impulse load.

8.3.1   Governing differential equations

In this work, as mentioned before, the governing wave equations for a MWNT are obtained using simplified Flügge equations for elastic shells. They are modified to take into account the interlayer coupling given by the vdW forces. For any arbitrary k^{th} wall, the three equations are as follows,

$$
\frac{\partial^2 u_k}{\partial x^2} + \frac{1}{2r_k^2}(1-\nu)\frac{\partial^2 u_k}{\partial \theta^2} + \frac{1}{2r_k}(1+\nu)\frac{\partial^2 v_k}{\partial x \partial \theta} + \frac{\nu}{r_k}\frac{\partial w_k}{\partial x}
$$
$$
+(1-\nu^2)\frac{D}{Ehr_k^2}\left[\frac{1}{2r_k^2}(1-\nu)\frac{\partial^2 u_k}{\partial \theta^2} - r_k\frac{\partial^3 w}{\partial x^3} + \frac{(1-\nu)}{2r_k}\frac{\partial^3 w}{\partial x \partial^2 \theta}\right]
$$
$$
= \frac{\rho h}{Eh}(1-\nu^2)\frac{\partial^2 u_k}{\partial t^2} \tag{8.1}
$$

$$
\frac{(1+\nu)}{2r_k}\frac{\partial^2 u_k}{\partial x \partial \theta} + \frac{(1-\nu)}{2}\frac{\partial^2 v_k}{\partial x^2} + \frac{1}{r_k^2}\frac{\partial^2 v_k}{\partial \theta^2} + \frac{1}{r_k^2}\frac{\partial w_k}{\partial \theta}
$$
$$
+(1-\nu^2)\frac{D}{Ehr_k^2}\left[\frac{3(1-\nu)}{2}\frac{\partial^2 v_k}{\partial x^2} - \frac{(3-\nu)}{2}\frac{\partial^3 w_k}{\partial x^2 \partial \theta}\right]
$$
$$
= \frac{\rho h}{Eh}(1-\nu^2)\frac{\partial^2 v_k}{\partial t^2} \tag{8.2}
$$

$$
\frac{\nu}{r_k}\frac{\partial u_k}{\partial x} + \frac{1}{r_k^2}\frac{\partial v_k}{\partial \theta} + \frac{w_k}{r_k^2} + (1-\nu^2)\frac{D}{Ehr_k^2}\left[r_k^2\frac{\partial^4 w_k}{\partial x^4} + 2\frac{\partial^4 w_k}{\partial x^2 \partial \theta^2} + \frac{1}{r_k^2}\frac{\partial^4 w_k}{\partial \theta^4}\right.
$$
$$
\left. -r_k\frac{\partial^3 u_k}{\partial x^3} + \frac{(1-\nu)}{2r_k}\frac{\partial^3 u_k}{\partial x \partial \theta^2} - \frac{(3-\nu)}{2}\frac{\partial^3 v_k}{\partial x^2 \partial \theta} + \frac{w_k}{r_k^2} + \frac{2}{r_k^2}\frac{\partial^2 w_k}{\partial \theta^2}\right]
$$
$$
= -\frac{1}{Eh}(1-\nu^2)\left[\rho h\frac{\partial^2 w_k}{\partial t^2} + p_k\right] \tag{8.3}
$$

where x and θ are the axial and circumferential dimensions respectively, while r_k is the radius of the kth tube and t is the time. u_k, v_k, and w_k are the displacements in axial, circumferential, and radial directions respectively for the kth wall. D is the effective bending stiffness, E is the Young's modulus,

ρ is the mass density, and ν is the Poisson's ratio and h is the thickness of each wall assumed for continuum modeling. The different values of these parameters are discussed in detail in Section 8.4 on numerical experiments. As said before, the governing equation corresponding to the radial displacement w_k given by Equation 8.3 for the kth is coupled through the term p_k which denotes the internal pressure resulting from vdW force and is given as

$$p_1 = c(w_2 - w_1) \tag{8.4}$$

$$p_k = c(w_{k+1} - w_k) - c\frac{r_{k-1}}{r_k}(w_k - w_{k-1}) \text{ for } k = 2, 3, \ldots, N-1 \tag{8.5}$$

$$p_N = -c\frac{r_{N-1}}{r_N}(w_N - w_{N-1}) \tag{8.6}$$

Here, c is the vdW interaction coefficient and is taken as

$$c = \frac{320 \times \text{erg/cm}^2}{0.16 d^2} \quad (d = 1.42 \times 10^{-8} \text{ cm})$$

For WSFE formulation, the associated force boundary conditions are required and they are derived as follows [28]. The resultant forces for the kth are given as

$$N_{\theta\theta} = E\left[\frac{1}{r_k}\frac{\partial v_k}{\partial \theta} - \frac{w_k}{r_k} + \nu\frac{\partial u_k}{\partial x}\right] \qquad N_{\theta x} = \frac{1}{2}(1-\nu)E\left[\frac{1}{r_k}\frac{\partial u_k}{\partial \theta} + \frac{\partial v_k}{\partial x}\right] \tag{8.7}$$

Similarly, the resultant moments are given as

$$M_{\theta\theta} = D\left[\frac{1}{r_k^2}\frac{\partial^2 w_k}{\partial \theta^2} + \nu\frac{\partial^2 w_k}{\partial x^2} + \frac{1}{r_k^2}\frac{\partial v_k}{\partial \theta}\right]$$

$$M_{\theta x} = \frac{1}{2r_k}D(1-\nu)\frac{\partial}{\partial x}\left[\frac{v_k}{r_k} + 2\frac{\partial w_k}{\partial \theta}\right] \tag{8.8}$$

From the resultant forces and moments given by Equations 8.7 and 8.8 respectively, the forces and moments at the boundary can be written as

$$Q_u = N_{\theta x} \tag{8.9}$$

$$Q_v = N_{\theta\theta} + \frac{1}{r_k}M_{\theta\theta} \tag{8.10}$$

$$Q_w = -\frac{1}{r_k}\frac{\partial M_{\theta\theta}}{\partial \theta} - 2\frac{\partial M_{\theta x}}{\partial x} \tag{8.11}$$

$$Q_m = M_{\theta\theta} \tag{8.12}$$

The governing equations (Equations 8.1 to 8.3) and the force boundary conditions (Equations 8.9 to 8.12) obtained using the Flügge's theory have variation in both axial (x) and circumferential (θ) directions. However, in this work, the WSFE model of the MWNT is developed considering variation only in θ direction. The reduced governing equations are of the following form,

$$\frac{1}{2r_k^2}(1-\nu)\left[1+(1-\nu^2)\frac{D}{Ehr_k^2}\right]\frac{\partial^2 u_k}{\partial\theta^2} = \frac{\rho h}{Eh}(1-\nu^2)\frac{\partial^2 u_k}{\partial t^2} \tag{8.13}$$

$$\frac{1}{r_k^2}\frac{\partial^2 v_k}{\partial\theta^2} + \frac{1}{r_k}\frac{\partial w_k}{\partial\theta} = \frac{\rho h}{Eh}(1-\nu^2)\frac{\partial^2 v_k}{\partial t^2} \tag{8.14}$$

$$\frac{1}{r_k^2}\frac{\partial v_k}{\partial\theta} + \frac{1}{r_k^2}\left[1+(1-\nu^2)\frac{D}{Ehr_k^2}\right]w_k + (1-\nu^2)\frac{D}{Ehr_k^4}\left[\frac{\partial^4 w_k}{\partial\theta^4} + 2\frac{\partial^2 w_k}{\partial\theta^2}\right]$$

$$= -\frac{1}{Eh}(1-\nu^2)\left[\rho h\frac{\partial^2 w_k}{\partial t^2} + p_k\right] \tag{8.15}$$

Thus, it can be seen from Equations 8.13 to 8.15 that the reduced equations are coupled only for the radial and circumferential motions, while the axial motion is uncoupled. The associated reduced boundary conditions are

$$Q_u = \frac{Eh}{2r_k(1+\nu)}\frac{\partial u_k}{\partial\theta} \tag{8.16}$$

$$Q_v = \left[\frac{Eh}{r_k(1-\nu^2)} + \frac{D}{r_k^3}\right]\left(\frac{\partial v_k}{\partial\theta} + \frac{\partial^2 w_k}{\partial\theta^2}\right) \tag{8.17}$$

$$Q_w = \frac{D}{r_k^4}\left(\frac{\partial^2 v_k}{\partial\theta^2} + \frac{\partial^3 w_k}{\partial\theta^3}\right) \tag{8.18}$$

$$Q_m = \frac{D}{r_k^2}\left(\frac{\partial^2 w_k}{\partial\theta^2} + \frac{\partial v_k}{\partial\theta}\right) \tag{8.19}$$

These reduced partial differential wave equations are used to formulate the WSFE model for MWNT in the next subsection.

8.3.2 Spectral finite element formulation

The governing equations 8.13 to 8.15 and the associated boundary conditions 8.16 to 8.19 are first reduced to ODEs using Daubechies scaling function as explained in the earlier chapters. The decoupled reduced ODEs corresponding to Equations 8.13 to 8.15 are as follows, with l varying from 0 to $n-1$

$$\frac{1}{2r_k^2}(1-\nu)\left[1+(1-\nu^2)\frac{D}{Ehr_k^2}\right]\frac{d^2\widehat{u}_l}{d\theta^2} = -\gamma_l^2\frac{\rho h}{Eh}(1-\nu^2)\widehat{u}_l \tag{8.20}$$

$$\frac{1}{r^2}\frac{d^2\widehat{v}_l}{d\theta^2} + \frac{1}{r}\frac{d\widehat{w}_l}{d\theta} = -\gamma_l^2\frac{\rho h}{Eh}(1-\nu^2)\widehat{v}_l \tag{8.21}$$

$$\frac{1}{r^2}\frac{d\widehat{v}_l}{d\theta} + \frac{1}{r^2}\left[1+(1-\nu^2)\frac{D}{Ehr^2}\right]\widehat{w}_l + (1-\nu^2)\frac{D}{Ehr^4}\left[\frac{d^4\widehat{w}_l}{d\theta^4} + 2\frac{d^2\widehat{w}_l}{d\theta^2}\right]$$

$$= -\frac{1}{Eh}(1-\nu^2)\left[-\gamma_l^2\rho h\widehat{w}_l + \widehat{p}_l\right] \tag{8.22}$$

where the notations carry their defined meaning.

Similarly, the boundary conditions given by Equations 8.16 to 8.19 are transformed as

$$\widehat{Q}_{ul} = \frac{Eh}{2r(1+\nu)} \frac{d\widehat{u}_l}{d\theta} \tag{8.23}$$

$$\widehat{Q}_{vl} = \left[\frac{Eh}{r(1-\nu^2)} + \frac{D}{r^3} \right] \left(\frac{d\widehat{v}_l}{d\theta} + \frac{d^2\widehat{w}_l}{d\theta^2} \right) \tag{8.24}$$

$$\widehat{Q}_{wl} = \frac{D}{r^4} \left(\frac{d^2\widehat{v}_l}{d\theta^2} + \frac{d^3\widehat{w}_l}{d\theta^3} \right) \tag{8.25}$$

$$\widehat{Q}_{ml} = \frac{D}{r^2} \left(\frac{d^2\widehat{w}_l}{d\theta^2} + \frac{d\widehat{v}_l}{d\theta} \right) \tag{8.26}$$

The spectral finite element has four degrees of freedom per node which are $\widehat{u}_l$, $\widehat{v}_l$, $\widehat{w}_l$, and $\frac{d\widehat{w}_l}{d\theta}$. The exact interpolating functions for an element of circumferential length $s_e = r\theta_e$, obtained by solving Equations 8.20 to 8.22 respectively, are

$$\{\widehat{u}(\theta),\ \widehat{v}(\theta),\ \widehat{w}(\theta)\}^T = [\mathbf{R}][\mathbf{\Theta}]\{\mathbf{a}\} \tag{8.27}$$

where $[\mathbf{\Theta}]$ is a diagonal matrix with the diagonal terms $[e^{-k_1\theta},\ e^{-k_1(\theta_e-\theta)},\ e^{-k_2\theta},\ e^{-k_2(\theta_e-\theta)},\ e^{-k_3\theta},\ e^{-k_3(\theta_e-\theta)},\ e^{-k_4\theta},\ e^{-k_4(\theta_e-\theta)}]$ and $[\mathbf{R}]$ is a 3×8 amplitude ratio matrix for each set of k_1, k_2, k_3, and k_4. Next, following the procedure of WSFE formulation, the exact dynamic stiffness matrix is obtained relating the transformed nodal displacements $\{\widehat{\mathbf{u}}^e\}$ with the transformed nodal forces $\{\widehat{\mathbf{F}}^e\}$. Here, $\{\widehat{\mathbf{u}}^e\} = \{\widehat{u}_1\ \widehat{w}_1\ d\widehat{w}_1/d\theta\ \widehat{u}_2\ \widehat{w}_2\ d\widehat{w}_2/d\theta\}$ and $\widehat{u}_1 \equiv \widehat{u}(0)$, $\widehat{v}_1 \equiv \widehat{v}(0)$, $\widehat{w}_1 \equiv \widehat{w}(0)$, $d\widehat{w}_1/dx \equiv d\widehat{w}(0)/dx$ and $\widehat{u}_2 \equiv \widehat{u}(\theta_e)$, $\widehat{v}_2 \equiv \widehat{v}(\theta_e)$, $\widehat{w}_2 \equiv \widehat{w}(\theta_e)$, $d\widehat{w}_2/d\theta \equiv d\widehat{w}/d\theta(\theta_e)$ and $\{\widehat{\mathbf{F}}^e\} = \{\widehat{Q}_{u1}\ \widehat{Q}_{v1}\ \widehat{Q}_{w1}\ \widehat{Q}_{m1}\ \widehat{Q}_{u2}\ \widehat{Q}_{v2}\ \widehat{Q}_{w2}\ \widehat{Q}_{m2}\}$ and $\widehat{Q}_{u1} \equiv \widehat{Q}_u(0)$, $\widehat{Q}_{v1} \equiv \widehat{Q}_v(0)$, $\widehat{Q}_{w1} \equiv \widehat{Q}_w(0)$, $\widehat{Q}_{m1} \equiv \widehat{Q}_m(0)$ and $\widehat{Q}_{u2} \equiv \widehat{Q}_u(\theta_e)$, $\widehat{Q}_{v2} \equiv \widehat{Q}_v(\theta_e)$, $\widehat{Q}_{w2} \equiv \widehat{Q}_w(\theta_e)$, $\widehat{Q}_{m2} \equiv \widehat{Q}_m(\theta_e)$.

The assembly of the elemental dynamic stiffness matrix to obtain the global stiffness matrix of the MWNT is explained through the example of a three walled carbon nanotube shown in Figure 8.6. The outermost wall of the MWNT is divided into two elements which are connected through two nodes using compatibility of displacements at the connecting nodes.

8.4   Frequency domain analysis

First, the wave properties are studied in the frequency domains. These include wavenumbers and group speed. The wavenumbers are studied for single-, double-, and three-walled carbon nanotubes with a fixed radius of the innermost tube. The thickness of each tube is taken as 0.34 nm, unless oth-

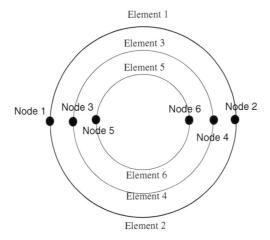

FIGURE 8.6: Schematic of the WSFE model of a three-walled carbon nanotube

erwise mentioned. The group speeds are obtained for a single-walled carbon nanotube with different radii.

The bulk material properties are Young modulus, $Eh = 360$ J/m^2 [102], mass density, $\rho h = 2270 \times 0.34$ kg/m^3, and Poisson's ratio, $\nu = 0.2$ [98]. In particular these parameters are not dependent on the definition of thickness $2h$ [102].

In Figures 8.7(a) to (c), the normalized wavenumbers $K_\theta R$ are presented for single-, double-, and three-walled carbon nanotubes respectively. The bold line shows the real part of the wavenumbers and the dotted line shows the imaginary part. In all the three cases, the innermost radius is $R = 2.0$ nm. A SWNT has three wave modes, namely, axial, circumferential, and radial. Correspondingly, double-, and three-walled nanotubes have six and nine wave modes, three for each wall. It can be seen from Figure 8.7(a) that the axial wavenumbers vary linearly with the frequency. This implies that the axial mode is non-dispersive in nature, i.e., the wave speeds do not vary with the frequency. As a result, the axial waves do not change their shapes as they propagate. The axial mode is purely real and hence does not dissipate as it propagates. On the other hand, the radial and the circumferential modes are dispersive in nature, or their shapes change as they propagate. In addition, both these modes have considerable imaginary part, which implies that these modes are dissipative in nature. From Figure 8.7(a), it can be observed that the circumferential mode has a frequency band within which it does not propagate. This can be considered as a cut-off frequency band within which the circumferential wave mode does not propagate. The radial mode, however, does not have such cut-off frequency band. It is dispersive in nature similar to the circumferential mode. As mentioned earlier, Figures 8.7(b) and (c) show the wavenumbers for two- and three-walled carbon nanotubes. The two-walled nanotube has a total of six wave modes, three for each wall. Similarly, the three-walled nanotube has nine wave modes. In Figures 8.7(b) and (c), similar patterns of the wave modes are observed. For the outer walls, however, the width of cut-off frequency band for the circumferential mode decreases. Again the magnitude of the wavenumbers for the outer walls increases with respect to the inner wall.

The group speeds of circumferential and radial wave modes of a SWNT are plotted in Figures 8.8(a) and (b), respectively for radius $R = 0.678$ nm and 2.0 nm. It can be seen that the group speed for the circumferential mode is zero within the cut-off frequency band. After this frequency band, the group speed is nearly constant with the variation of frequency. Apart from this, this constant value of circumferential wave velocity is independent of the radius, as it is observed from Figures 8.7(a) and (b). This is in tune with that observed in reference [48], where an atomistic simulation is done to obtain the wave velocities in SWNT. From Figure 8.8(b) it can be observed that the cut-off frequency band for SWNT with $R = 2.0$ nm is much smaller compared to that for SWNT with $R = 0.678$. In addition, this cut-off band occurs at a much lower frequency range for $R = 2.0$ nm than $R = 0.678$ nm. The radial mode, however, has a non-zero group speed over the entire frequency range. The group speed for the radial mode is nearly equal to the group speed of the circumferential mode at higher range of frequencies, while there is considerable difference at lower frequencies, approximately below 20 THz. The axial wave velocity does not vary with frequency and also is invariant of the radius of SWNT. The axial wave velocity predicted by the present method is 14.0 km/s. The corresponding velocity predicted by Yu et al. [105] is 17.5 km/s.

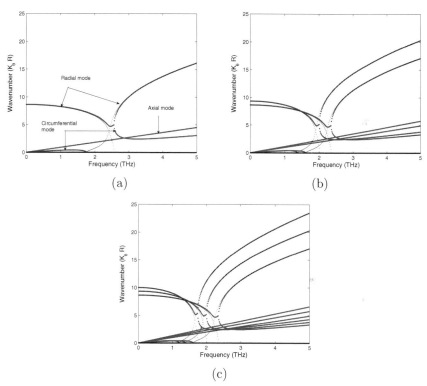

(a)   (b)

(c)

FIGURE 8.7:   Normalized wavenumber $(K_\theta R)$ for (a) single-walled, (b) double-walled, and (c) three-walled carbon nanotubes

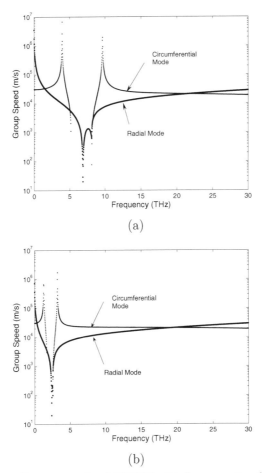

FIGURE 8.8: Group speeds of SWNT with inner radius (a) 0.678 nm and (b) 2.0 nm

8.5   Time domain analysis

Next, the time domain responses of SWNT and MWNT are simulated. The wave propagation results from short duration impulse load of frequency content 44 THz and duration 50 ps. These loads are applied at diametrically opposite nodal point on the CNT walls (see Figure 8.6). The axial, circumferential, and radial wave propagations are studied for the impulse load being applied in the corresponding directions. In Figures 8.9(a) to (d), the snapshots of radial wave propagation in a SWNT with inner radius 2.0 nm are presented for time instances 150 ps, 200 ps, 250 ps, and 300 ps respectively.

The amplitudes of the waves plotted are normalized for better interpretation. Similarly, in Figures 8.10(a) to (d), the snapshots of circumferential wave propagation are plotted for time instances 150 ps, 200 ps, 250 ps, and 300 ps respectively. Even here, the loading is similar to that for the previous case except that the load is applied in circumferential direction.

In Figure 8.11, the first five natural frequencies corresponding to the axial mode of vibration of SWNT are plotted for different values of the innermost radius, R. The frequencies are obtained from the frequency response function (FRF) of the axial displacements measured at the nodes (refer to Figure 8.6). As mentioned earlier, the loading is the broadband impulse load applied at the nodes along the axial directions. It can be observed that the frequencies of all the five modes vary linearly with R and the slope of variation increases for higher modes. Here, the frequencies are within a range of 2.0 to 12.0 THz and decrease with increase in R. Similarly, Figure 8.12 presents the frequencies of the first five circumferential vibration modes of SWNT for varying innermost radius, R. Even here, the frequencies for all the modes vary linearly with R. The slope increases for the higher modes, while the variation is almost constant for the first mode. The loading condition is similar to that of the previous example except that they are applied in the circumferential direction. The frequencies are obtained from the FRF of the circumferential displacements measured at the nodal points. The frequencies of these five modes vary within 3.0 to 17.0 THz which is slightly higher than that for the axial modes. Finally, the natural frequencies of the first five radial modes are plotted in Figure 8.13 for different values of R. The impulse load described before is applied at the nodes of the WSFE model (refer Figure 8.6) for the SWNT in radial direction and the frequencies are obtained from the FRF of the radial displacements at the nodal points. For the radial modes, the range within which the frequencies vary is 1.0 to 4.0 THz and is much smaller as compared to the two previous cases. Here again, the frequencies vary linearly with R, but, unlike axial and circumferential modes, the slope of the variations is nearly equal for all the five modes. It should be mentioned that the radial and circumferential vibrations are coupled though the coupled wave propagation is not studied here.

In Figure 8.14, the axial wave velocities of the walls of a three-walled carbon nanotubes are plotted. The wave responses result from the unit amplitude impulse load applied at the nodes 1 to 6 as shown in Figure 8.6. It can be seen that the wave velocity is maximum for the innermost wall (referred as wall 1 in the figure) and increases for the outer walls. However, the wave velocity patterns match for all the three walls. For the three-walled carbon nanotubes configuration considered, the innermost radius is 2.0 nm and the wall thickness is 0.34 nm as mentioned earlier. Similar wave responses are presented in Figure 8.15, but for circumferential wave propagation. Even here, the wave velocity is highest for the innermost wall and increases for the outer walls, while the time history of the velocities follows a similar trend. Finally, Figure 8.16 presents the wave velocities of the three-walled carbon

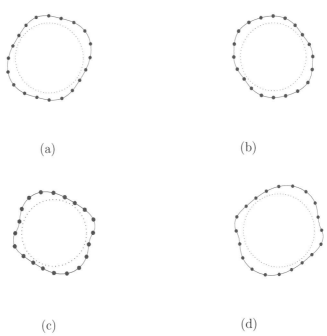

(a)   (b)

(c)   (d)

FIGURE 8.9:   Snapshots of radial wave propagation in SWNT with inner radius $R = 2.0$ nm at time instances (a) 150 ps, (b) 200 ps, (c) 250 ps, and (d) 300 ps

nanotubes corresponding to the radial wave propagation. However, here there are not much differences in the wave velocities of the three walls but the initial amplitudes vary considerably. This is expected as the radial stiffness of the walls is different due to difference in the radius.

In conclusion it can be said that this chapter studies the vibration and wave properties of carbon nanotubes, both single- and multi-walled through continuum beam and shell models. The one- and two-dimensional WSFE schemes developed in the last three chapters are implemented here. As mentioned earlier, WSFE is found to be a tool ideally suited for this purpose. The dynamical properties of CNTs observed from these analyses have been validated with experimental and atomistic simulation results available in literature. A good corroboration is observed in all the cases. Following this understanding, an attempt is made to explore the dynamic properties of nano-composites referring to CNT reinforced matrix material in the next chapter.

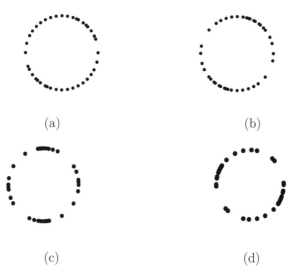

(a) (b)

(c) (d)

FIGURE 8.10: Snapshots of circumferential wave propagation in SWNT with inner radius $R = 2.0$ nm at time instances (a) 150 ps, (b) 200 ps, (c) 250 ps, and (d) 300 ps

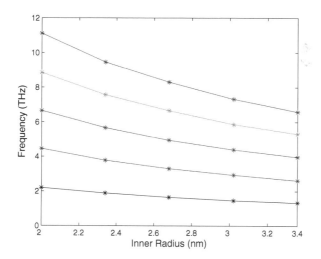

FIGURE 8.11: Natural frequencies of the first five axial modes of a SWNT with varying inner radius R

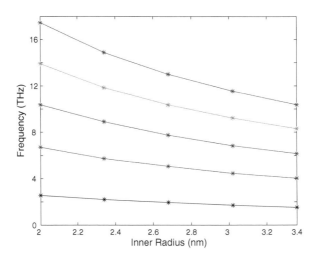

FIGURE 8.12: Natural frequencies of the first five circumferential modes of a SWNT with varying inner radius R

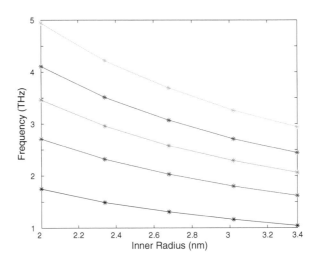

FIGURE 8.13: Natural frequencies of the first five radial modes of a SWNT with varying inner radius R

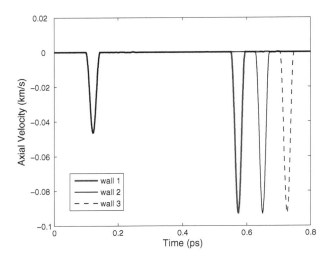

FIGURE 8.14:   Axial wave propagation in a three-walled carbon nanotube of innermost radius $R = 2.0$ nm

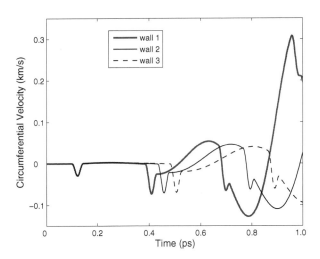

FIGURE 8.15:   Circumferential wave propagation in a three-walled carbon nanotube of innermost radius $R = 2.0$ nm

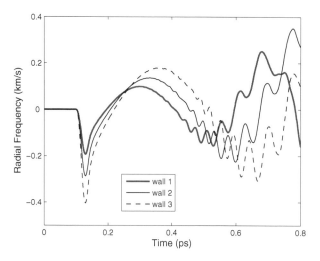

FIGURE 8.16:   Radial wave propagation in a three-walled carbon nanotube of innermost radius $R = 2.0$ nm

Chapter 9

Vibration and Wave Propagation in Nano-Composites

As demonstrated in the last chapter, WSFE is found to be an effective computational tool for dynamic analysis of CNTs, both single- and multi-walled. Different continuum shell models were implemented for the analysis. CNTs are however mostly used in conjunction with matrix materials in form of nano-composites. Such nano-composites possess immense potential as structural material with extraordinary mechanical properties and functionalities apart from their use in other device applications like sensors and actuators. Though understanding the mechanical properties of CNTs alone is the basic requirement, high fidelity modeling of CNT-based composites is also essential due to several additional complexities associated with such nano-composites. This chapter explains continuum modeling of CNT embedded composites generally referred to as nano-composite using WSFE scheme. The models are then used for wave propagation analysis of such structures.

9.1 Introduction: nano-composites

The high strength of CNTs, as said before, is a powerful motivation for their use as reinforcing materials to form high strength, lightweight composites with functional properties. Embedding CNTs will help to enhance the damping characteristics, increase buckling capacity, and improve resistance to higher temperature. In addition, for other device applications like sensors and actuators, CNTs are always embedded in matrix materials, mostly polymer matrix. One main advantage of nano-composite is that it does not require precise alignment of CNTs. Thus, it is expected that the boarder applications of CNTs will evolve in the form of nano-composites [92]. The properties of nano-composites are however not yet well understood and a considerable work needs to be done to gain a good knowledge about their behavior [94].

The two main problems associated with the study and modeling of nano-composites are as follows. First, the scale difference, i.e., the CNTs interact with the matrix at nano-scale while the actual nano-composite is in millimeter to meter range. Traditional modeling techniques with single time and space scales cannot handle such structures and a multi-scale modeling scheme is required to break the boundaries of scales [75]. Second, the strength of the nano-composite is governed by the stress transfer between the CNT and the surrounding matrix. In general, perfect bonding does not occur between these and the amount of stress transfer at the interface can be obtained only through experimental study or atomistic simulations.

Since controlled experiments at the nano-scale are difficult and with atomistic simulation being computationally expensive, continuum modeling can prove to be very valuable in the advancement of nano-scale structures but it needs further improvement and development. Nano-composites are generally analyzed considering a representative volume element (RVE) consisting of CNTs either single- or multi-walled surrounded by matrix material. Several beam and shell models have been implemented to model such RVEs. To explain a few, the authors [64] formulated a RVE with MWNT embedded in different matrix materials and is modeled as the beam using layerwise third order shear deformation theory. This model accounts for partial shear stress transfer. Later, the authors [67] developed a continuum axisymmetric shell model-based RVE considering pull-out and partial stress transfer. An RVE of a composite reinforced with a capped SWNT has been modeled using a modified shear-lag model [34]. The interfacial shear strength in a polymer composite reinforced has been modeled with a SWNT using a modified Kelly-Tyson approach [97]. A molecular structural mechanics-based continuum model has been derived [47] to model nano-composites taking into account the interfacial stress transfer.

In this chapter, a beam and axisymmetric shell model of RVE consisting of CNT embedded in matrix material developed by the authors [64, 67] as mentioned earlier are explained in detail. The modeling is done using one- and two-dimensional WSFE and used to study dynamic properties. In the first case, the nano-composite RVEs consisting of MWNT embedded in polymer, metal, and ceramic matrices are modeled using third order layer-wise shear deformation theory [73]. Each wall of the MWNT is modeled as an individual beam coupled through a distributed spring throughout the length of the beam. The stiffness of the spring depends on the inter-layer van der Waals force [86]. The layer-wise theory is adopted to account for partial interfacial shear stress transfer. In the second case, a RVE of SWNT-polymer composite is developed. The RVE is modeled as an axisymmetric finite length cylinder with coupled axial and radial degrees of freedom. The imperfect bonding between the SWNT and the matrix resulting in the phenomenon of CNT pull-out is modeled considering non-uniform axial strains at the SWNT-polymer interface. Apart from this, the effect of partial shear stress transfer is also modeled. The effects are included in the model as constraints which are imposed using the penalty matrix method [18]. These models are described elaborately in the following Sections.

9.2 Beam model of MWNT embedded nano-composite

Modeling of P-walled MWNT-composite beam as coupled multiple-elastic beam [86] as shown in Figure 9.1(a), based on third order layerwise shear deformation theory, results in a set of $2P$ governing coupled PDEs. Each CNT wall is associated with two PDEs corresponding to transverse and shear displacements. Since the matrix is bonded only to the outermost CNT, the layer-wise theory is adopted only for deriving the two PDEs related to matrix and outermost wall. Thus, two sets of displacement fields are used for the matrix and CNT wall, and is given as

$$u^m = -z\frac{\partial w^1}{\partial x} + (z + c_1^m z^2 + c_2^m z^3)\theta^1 \tag{9.1}$$

$$w^m = w^1 \tag{9.2}$$

for matrix and similarly for the outermost CNT wall,

$$u^{c1} = -z\frac{\partial w^1}{\partial x} + (z + c_1^{c1} z^2 + c_2^{c1} z^3)\theta^1 \tag{9.3}$$

$$w^{c1} = w^1 \tag{9.4}$$

where w^1 is the displacement and θ^1 is the rotation due shear in z direction as shown in Figure 9.1(b). c_1^m, c_2^m and c_1^{c1}, c_2^{c1} are the constants that need to be determined.

Strain displacement relations are derived for the matrix using Equations 9.1 to 9.2

$$\varepsilon_{xx}^m = \frac{\partial u^m}{\partial x} = -z\frac{\partial^2 w^1}{\partial x^2} + (z + c_1^m z^2 + c_2^m z^3)\frac{\partial \theta^1}{\partial x} \tag{9.5}$$

$$\gamma_{xz}^m = \frac{\partial u^m}{\partial z} + \frac{\partial w^m}{\partial x} = (1 + 2c_1^m z + 3c_2^m z^2)\theta^1 \tag{9.6}$$

and similarly for outermost CNT using Equations 9.3 to 9.4

$$\varepsilon_{xx}^{c1} = \frac{\partial u^{c1}}{\partial x} = -z\frac{\partial^2 w^1}{\partial x^2} + (z + c_1^{c1} z^2 + c_2^{c1} z^3)\frac{\partial \theta^1}{\partial x} \tag{9.7}$$

$$\gamma_{xz}^{c1} = \frac{\partial u^{c1}}{\partial z} + \frac{\partial w^{c1}}{\partial x} = (1 + 2c_1^{c1} z + 3c_2^{c1} z^2)\theta^1 \tag{9.8}$$

The constants c_1^m and c_2^m are determined by assuming that there are no distributed shear loads on the top and bottom surfaces of the beam, or

$$\tau_{xz}^m = G^m \gamma_{xz}^m = 0 \quad \text{for} \quad z = \pm h \tag{9.9}$$

where G^m is the shear modulus of the matrix and h is half the beam depth as in Figure 9.1(b). Solving Equation 9.9, c_1^m and c_2^m are obtained as

$$c_1^m = 0 \quad \text{and} \quad c_2^m = -\frac{1}{3h^2} \tag{9.10}$$

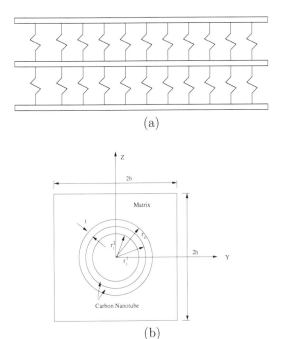

FIGURE 9.1:   (a) Distributed spring model of coupling between the carbon nanotubes; (b) beam cross-section with embedded double wall carbon nanotube

The other two constants c_1^c and c_2^c are derived by assuming continuous displacement and given fraction of shear stress transfer at the CNT matrix interface.

$$u^{c1} = u^m \quad \text{for} \quad z = r_o \sin \psi \ \forall \ \psi \ \in \ (0, \ 2\pi) \tag{9.11}$$

$$\text{or,} \ \ c_1^{c1} + c_2^{c1} z = c_2^m \ \ \text{for} \ z = r_o \sin \psi \ \forall \ \psi \ \in \ (0, \ 2\pi) \tag{9.12}$$

and,

$$\tau_{xz}^{c1} = \alpha \tau_{xz}^m \ \text{for} \ z = r_o \sin \psi \ \forall \ \psi \ \in \ (0, \ 2\pi) \tag{9.13}$$

$$\text{or,} \ (1 + 2c_1^{c1}z + 3c_2^{c1}z^2) = \frac{\alpha G^m}{G^c}(1 + 3c_2^{c1}z^2)$$
$$\text{for} \ z = r_o \sin \psi \ \forall \ \psi \ \in \ (0, \ 2\pi) \tag{9.14}$$

$$\text{or,} \ \ 2c_1^{c1}z + 3c_2^{c1}z^2 = -(1 - \kappa) + 3\kappa c_2^m z^2 \quad \kappa = \frac{\alpha G^m}{G^c} \tag{9.15}$$

where r_o is the radius and G^c is the shear modulus of the outermost CNT and α is the given fraction of interfacial shear stress transfer. Note that $0 \ \leq \ \alpha \ \leq \ 1$, $\alpha = 1$ denotes full transfer and $\alpha = 0$ denote no transfer of interfacial shear stress.

By solving Equations 9.12 and 9.15, c_1^{c1} and c_2^{c1} are obtained as

$$c_1^{c1} = -(1 - \kappa)\frac{1}{z^2} + (3\kappa - 2)c_2^m \tag{9.16}$$

$$c_2^{c1} = 3(1 - \kappa)c_2^m z + (1 - \kappa)\frac{1}{z} \ \text{for} \ z = r_o \sin \psi \ \forall \ \psi \ \in \ (0, \ 2\pi) \tag{9.17}$$

By substituting, $\theta^1 = \phi^1 + \frac{\partial w^1}{\partial x}$ where ϕ^1 is the total rotation at $z = 0$, in Equations 9.5 to 9.8, they can be written as

$$\varepsilon_{xx}^m = (c_1^m z^2 + c_2^m z^3)\frac{\partial^2 w^1}{\partial x^2} + (z + c_1^m z^2 + c_2^m z^3)\frac{\partial \phi^1}{\partial x} \tag{9.18}$$

$$\gamma_{xz}^m = (1 + 2c_1^m z + 3c_2^m z^2)(\phi^1 + \frac{\partial w^1}{\partial x}) \tag{9.19}$$

and

$$\varepsilon_{xx}^{c1} = (c_1^{c1} z^2 + c_2^{c1} z^3)\frac{\partial^2 w^1}{\partial x^2} + (z + c_1^{c1} z^2 + c_2^{c1} z^3)\frac{\partial \phi^1}{\partial x} \tag{9.20}$$

$$\gamma_{xz}^{c1} = (1 + 2c_1^{c1} z + 3c_2^{c1} z^2)(\phi^1 + \frac{\partial w^1}{\partial x}) \tag{9.21}$$

Total strain and kinetic energies for the matrix and outermost CNT are calculated as

$$U^1 = \frac{1}{2}\int_0^L \int_A (E^m \varepsilon_{xx}^{m\,2} + E^c \varepsilon_{xx}^{c1\,2} + G^m \gamma_{xz}^{m\,2} + E^c \gamma_{xz}^{c1\,2})dAdx \tag{9.22}$$

$$T^1 = \frac{1}{2}\int_0^L \int_A (\rho^m(\dot{u}^{m^2} + \dot{w}^{m^2}) + \rho^c(\dot{u}^{c1^2} + \dot{w}^{c1^2}))dAdx \tag{9.23}$$

For the remaining $P-1$ walls of the P walled MWNT, the displacement fields are similar to those given by Equations 9.3 and 9.4 and can be written as

$$u^{cp} = -z\frac{\partial w^p}{\partial x} + (z + c_1^{cp}z^2 + c_2^{cp}z^3)\theta^p \tag{9.24}$$

$$w^{cp} = w^p \quad \text{for} \quad p = 2, \, 3, \ldots P \tag{9.25}$$

where w^p is the transverse displacement and θ^p is the rotation due to shear in z direction for pth CNT wall. Similarly, the strain displacement relations for these CNT walls can be written as (after using $\theta^p = \phi^p + \frac{\partial w^p}{\partial x}$)

$$\varepsilon_{xx}^{cp} = (c_1^{cp}z^2 + c_2^{cp}z^3)\frac{\partial^2 w^p}{\partial x^2} + (z + c_1^{cp}z^2 + c_2^{cp}z^3)\frac{\partial \phi^p}{\partial x} \tag{9.26}$$

$$\gamma_{xz}^{cp} = (1 + 2c_1^{cp}z + 3c_2^{cp}z^2)(\phi^p + \frac{\partial w^p}{\partial x}) \quad \text{for} \quad p = 2, \, 3, \ldots P \tag{9.27}$$

where ϕ^p is the total rotation in z direction at $z = 0$. c_1^{cp} and c_2^{cp} are constants similar to c_1^{cp} and c_2^{cp} respectively, as given in Equation 9.3. However, unlike the outermost CNT wall, these constants are derived by assuming that the shear loads at the free outer surface is zero or

$$\tau_{xz}^{cp} = G^c\gamma_{xz}^{cp} = 0 \quad \text{for} \quad z = \pm r_i^{(p-1)}\sin\psi, \, \forall \, \psi \, \in \, (0, \, \pi) \quad p = 2, \, 3, \ldots, P \tag{9.28}$$

where $r_i^{(p-1)}$ is the inner radius of $(p-1)$th CNT wall. From Equation 9.28, the constants c_1^{cp} and c_2^{cp} can be solved as

$$c_1^{cp} = 0 \quad \text{and} \quad c_2^{cp} = -\frac{1}{3z^2} \text{ for } z = r_i^{(p-1)}\sin\psi, \, \forall \, \psi \, \in \, (0, \, 2\pi)$$

$$p = 2, \, 3, \ldots, P \tag{9.29}$$

Thus the potential and kinetic energies for these CNT walls can be obtained as

$$U^P = \frac{1}{2}\int_0^L \int_A (E^c\varepsilon_{xx}^{cp\,2} + E^c\gamma_{xz}^{cp\,2})dAdx \tag{9.30}$$

$$T^P = \frac{1}{2}\int_0^L \int_A \rho^c(\dot{u}^{cp\,2} + \dot{w}^{cp\,2})dAdx \quad p = 2, \, 3, \ldots, P \tag{9.31}$$

Using the Hamilton's principle, the minimization of the above energies with respect to w^p and ϕ^p, $p = 1, \, 2, \, 3, \ldots, P$, gives $2P$ coupled governing PDEs as

$$I_0^1\ddot{w}^1 - I_{ff}^1\frac{\partial^2\ddot{w}^1}{\partial x^2} + I_{fs}^1\frac{\partial\ddot{\phi}^1}{\partial x} + K_{ff}^1\frac{\partial^4 w^1}{\partial x^4} - K_{sf}^1(\frac{\partial^2 w^1}{\partial x^2} + \frac{\partial\phi^1}{\partial x})$$

$$+K_{fs}^1\frac{\partial^3\phi^1}{\partial^3 x} = C_1[w^2 - w^1] \tag{9.32}$$

$$I_{ss}^1 \ddot{\phi}^1 + I_{fs}^1 \frac{\partial \ddot{w}^1}{\partial x} + K_{sf}^1 (\frac{\partial w^1}{\partial x} + \phi^1) - K_{fs}^1 \frac{\partial^3 w^1}{\partial^3 x} - K_{ss}^1 \frac{\partial^2 \phi^1}{\partial x^2} = 0 \qquad (9.33)$$

$$I_0^p \ddot{w}^p - I_{ff}^p \frac{\partial^2 \ddot{w}^p}{\partial x^2} + I_{fs}^p \frac{\partial \ddot{\phi}^p}{\partial x} + K_{ff}^p \frac{\partial^4 w^p}{\partial x^4} - K_{sf}^p (\frac{\partial^2 w^p}{\partial x^2} + \frac{\partial \phi^p}{\partial x}) + K_{fs}^p \frac{\partial^3 \phi^p}{\partial^3 x}$$
$$= C_p [w^{p+1} - w^p] - C_{p-1}[w^p - w^{p-1}] \quad p = 2, \, 3, \dots, P-1 \qquad (9.34)$$

$$I_{ss}^p \ddot{\phi}^p + I_{fs}^p \frac{\partial \ddot{w}^p}{\partial x} + K_{sf}^p (\frac{\partial w^p}{\partial x} + \phi^p) - K_{fs}^p \frac{\partial^3 w^p}{\partial^3 x} - K_{ss}^p \frac{\partial^2 \phi^p}{\partial x^2} = 0 \qquad (9.35)$$

$$I_0^P \ddot{w}^P - I_{ff}^P \frac{\partial^2 \ddot{w}^P}{\partial x^2} + I_{fs}^P \frac{\partial \ddot{\phi}^P}{\partial x} + K_{ff}^P \frac{\partial^4 w^P}{\partial x^4} - K_{sf}^P (\frac{\partial^2 w^P}{\partial x^2} + \frac{\partial \phi^P}{\partial x})$$
$$+ K_{fs}^P \frac{\partial^3 \phi^P}{\partial^3 x} = -C_{P-1}[w^P - w^{P-1}] \qquad (9.36)$$

$$I_{ss}^P \ddot{\phi}^P + I_{fs}^P \frac{\partial \ddot{w}^P}{\partial x} + K_{sf}^P (\frac{\partial w^P}{\partial x} + \phi^P) - K_{fs}^P \frac{\partial^3 w^P}{\partial^3 x} - K_{ss}^P \frac{\partial^2 \phi^P}{\partial x^2} = 0 \qquad (9.37)$$

It should be mentioned here that the governing differential equation corresponding to transverse displacement for pth CNT wall is coupled only with those for $(p-1)$th and $(p+1)$th walls. The inertial constants for the Equations 9.32 and 9.33 are

$$I_{ff}^1 = (c_1^{m 2} I_4^m + c_1^{c1 2} I_4^{c1}) + 2(c_1^m c_2^m I_5^m + c_1^{c1} c_2^{c1} I_5^{c1})$$
$$+ (c_2^{m 2} I_6^m + c_2^{c1 2} I_6^{c1}) \qquad (9.38)$$
$$I_{fs}^1 = I_{ff}^1 + (c_1^m I_3^m + c_1^{c1} I_3^{c1}) + (c_2^m I_4^m + c_2^{c1} I_4^{c1}) \qquad (9.39)$$
$$I_{ss}^1 = I_{fs}^1 + (I_2^m + I_2^{c1}) + (c_1^m I_3^m + c_1^{c1} I_3^{c1}) + (c_2^m I_4^m + c_2^{c1} I_4^{c1}) \quad (9.40)$$

and similarly, for Equations 9.33 and 9.37,

$$I_{ff}^p = c_1^{c1 2} I_4^{cp} + 2c_1^{cp} c_2^{cp} I_5^{cp} + c_2^{cp 2} I_6^{cp} \qquad (9.41)$$
$$I_{fs}^p = I_{ff}^p + c_1^{cp} I_3^{cp} + c_2^{cp} I_4^{cp} \qquad (9.42)$$
$$I_{ss}^p = I_{fs}^p + I_2^{cp} + c_1^{cp} I_3^{cp} + c_2^{cp} I_4^{cp} \quad \text{for} \quad p = 2, \, 3, \dots, P \qquad (9.43)$$

and

$$c_{()}^{()} I_l^{()} = \rho^{()} \int_A c_{()}^{()} z^l dA$$

where the superscripts are either m or cp, $p = 1, \, 2, \dots, P$ and the subscripts are either 1 or 2 as in Equations 9.38 to 9.43.

The stiffness constants for the Equations 9.32 and 9.33 are

$$K_{ff}^1 = (c_1^{m 2} F_{11}^m + c_1^{c1 2} F_{11}^{c1}) + 2(c_1^m c_2^m G_{11}^m + c_1^{c1} c_2^{c1} G_{11}^{c1})$$
$$+ (c_2^{m 2} H_{11}^m + c_2^{c1 2} H_{11}^{c1}) \qquad (9.44)$$

$$K_{fs}^1 = K_{ff}^1 + (c_1^m E_{11}^m + c_1^{c1} E_{11}^{c1}) + (c_2^m F_{11}^m + c_2^{c1} F_{11}^{c1}) \tag{9.45}$$

$$K_{sf}^1 = (A_{55}^m + A_{55}^{c1}) + 4(c_1^m B_{55}^m + c_1^{c1} B_{55}^{c1})$$

$$+[(4c_1^{m2} + 6c_2^m)D_{55}^m + (4c_1^{c1^2} + 6c_2^{c1})D_{55}^{c1}]$$

$$+12(c_1^m c_2^m E^m 55 + c_1^{c1} c_2^{c1} E^{c1} 55) + 9(c_2^{m2} F_{55}^m + c_2^{c1^2} F_{55}^{c1}) \tag{9.46}$$

$$K_{ss}^1 = K_{fs}^1 + (D_{11}^m + D_{11}^{c1}) + (c_1^m E_{11}^m + c_1^{c1} E_{11}^{c1})$$

$$+(c_2^m F_{11}^m + c_2^{c1} F_{11}^{c1}) \tag{9.47}$$

and similarly, for Equations 9.33 and 9.37,

$$K_{ff}^p = c_1^{cp2} F_{11}^{cp} + 2c_1^{cp} c_2^{cp} G_{11}^{cp} + c_2^{cp2} H_{11}^{cp} \tag{9.48}$$

$$K_{fs}^p = K_{ff}^p + c_1^{cp} E_{11}^{cp} + c_2^{cp} F_{11}^{cp} \tag{9.49}$$

$$K_{sf}^p = A_{55}^{cp} + 4c_1^{cp} B_{55}^{cp} + (4c_1^{cp2} + 6c_2^{cp})D_{55}^{cp} + 12c_1^{cp} c_2^{cp} E^{cp} 55$$

$$+9c_2^{cp2} F_{55}^{cp} \tag{9.50}$$

$$K_{ss}^p = K_{fs}^p + D_{11}^{cp} + c_1^{c1} E_{11}^{cp} + c_2^{cp} F_{11}^{cp} \quad \text{for} \quad p = 2, 3, \ldots, P \tag{9.51}$$

and

$$[A_{11}^{()}, c_{()}^{()} B_{11}^{()}, c_{()}^{()} D_{11}^{()}, c_{()}^{()} E_{11}^{()}, c_{()}^{()} F_{11}^{()}, c_{()}^{()} G_{11}^{()}, c_{()}^{()} H_{11}^{()}]$$

$$= E^{()} \int_A c_{()}^{()} [1, z, z^2, z^3, z^4, z^5, z^6] dA$$

$$[A_{55}^{()}, c_{()}^{()} B_{55}^{()}, c_{()}^{()} D_{55}^{()}, c_{()}^{()} E_{55}^{()}, c_{()}^{()} F_{55}^{()}, c_{()}^{()} G_{55}^{()}, c_{()}^{()} H_{55}^{()}]$$

$$= G^{()} \int_A c_{()}^{()} [1, z, z^2, z^3, z^4, z^5, z^6] dA$$

where similar to the inertial constants, the superscripts are either m or cp, $p = 1, 2, \ldots, P$ and the subscripts are either 1 or 2 as in Equations 9.44 to 9.51.

The interaction coefficients C_p, $p = 1, 2, \ldots, P - 1$, arising due to van der Waals interaction between any two adjacent walls, can be estimated approximately as [86]

$$C_p = \frac{400r_i^p}{0.16d^2} \quad \text{erg/cm}^2, \quad d = 0.142 \text{ nm}, \quad p = 1, 2, \ldots, P - 1 \tag{9.52}$$

where r_i^p is the inner radius of pth CNT wall.

The three force boundary conditions obtained by Hamilton's principle corresponding to w^p, $\frac{\partial w^p}{\partial x}$, and ϕ^p respectively are

$$-K_{ff}^p \frac{\partial^3 w^p}{\partial x^3} + K_{sf}^p \left(\frac{\partial w^p}{\partial x} + \phi^p\right) - K_{fs}^p \frac{\partial^2 \phi^p}{\partial x^2} = V \tag{9.53}$$

$$K_{ff}^p \frac{\partial^2 w^p}{\partial x^2} + K_{fs}^p \frac{\partial \phi^p}{\partial x} = Q \tag{9.54}$$

$$K_{fs}^p \frac{\partial^2 w^p}{\partial x^2} + K_{ss}^p \frac{\partial \phi^p}{\partial x} = M \quad p = 1, 2, \ldots, P \tag{9.55}$$

9.3 Spectral finite element formulation for MWNT embedded nano-composite beam

The reduced and the decoupled forms of the governing differential equations given by Equations 9.32 to 9.37 following Daubechies scaling function approximation, imposition of boundary conditions through wavelet extrapolation technique, and eigenvalue analysis as discussed in detail in the earlier chapters are as follows,

$$
-I_0^1 \gamma_j^2 \widehat{w}_j^1 + I_{ff}^1 \gamma_j^2 \frac{d^2 \widehat{w}_j^1}{dx^2} - I_{fs}^1 \gamma_j^2 \frac{d\widehat{\phi}_j^1}{dx} + K_{ff}^1 \frac{d^4 \widehat{w}_j^1}{dx^4} - K_{sf}^1 \left(\frac{d^2 \widehat{w}_j^1}{dx^2} + \frac{d\widehat{\phi}_j^1}{dx} \right)
$$

$$
+ K_{fs}^1 \frac{d^3 \widehat{\phi}_j^1}{dx^3} = C_1 \left[\widehat{w}_j^2 - \widehat{w}_j^1 \right] \tag{9.56}
$$

$$
-I_{ss}^1 \gamma_j^2 \widehat{\phi}_j^1 - I_{fs}^1 \gamma_j^2 \frac{d\widehat{w}_j^1}{dx} + K_{sf}^1 (\frac{d\widehat{w}_j^1}{dx} + \widehat{\phi}_j^1) - K_{fs}^1 \frac{d^3 \widehat{w}^1}{d^3 x}
$$

$$
- K_{ss}^1 \frac{d^2 \widehat{\phi}_j^1}{dx^2} = 0 \tag{9.57}
$$

$$
-I_0^p \gamma_j^2 \widehat{w}_j^p + I_{ff}^p \gamma_j^2 \frac{d^2 \widehat{w}_j^p}{dx^2} - I_{fs}^p \gamma_j^2 \frac{d\widehat{\phi}_j^p}{dx} + K_{ff}^p \frac{d^4 \widehat{w}_j^p}{dx^4} - K_{sf}^p \left(\frac{d^2 \widehat{w}_j^p}{dx^2} + \frac{d\widehat{\phi}_j^p}{dx} \right)
$$

$$
+ K_{fs}^p \frac{d^3 \widehat{\phi}_j^p}{dx^3} = C_p \left[\widehat{w}_j^{p+1} - \widehat{w}_j^p \right] - C_{p-1} \left[\widehat{w}_j^p - \widehat{w}_j^{p-1} \right] \quad p = 2 \text{ to } P-1 \tag{9.58}
$$

$$
-I_{ss}^p \gamma_j^2 \widehat{\phi}_j^p - I_{fs}^p \gamma_j^2 \frac{d\widehat{w}_j^p}{dx} + K_{sf}^p (\frac{d\widehat{w}_j^p}{dx} + \widehat{\phi}_j^p) - K_{fs}^p \frac{d^3 \widehat{w}^p}{d^3 x}
$$

$$
- K_{ss}^p \frac{d^2 \widehat{\phi}_j^p}{dx^2} = 0 \tag{9.59}
$$

$$
-I_0^P \gamma_j^2 \widehat{w}_j^P + I_{ff}^P \gamma_j^2 \frac{d^2 \widehat{w}_j^P}{dx^2} - I_{fs}^P \gamma_j^2 \frac{d\widehat{\phi}_j^P}{dx} + K_{ff}^P \frac{d^4 \widehat{w}_j^P}{dx^4} - K_{sf}^P \left(\frac{d^2 \widehat{w}_j^P}{dx^2} + \frac{d\widehat{\phi}_j^P}{dx} \right)
$$

$$
+ K_{fs}^P \frac{d^3 \widehat{\phi}_j^P}{dx^3} = -C_{P-1} \left[\widehat{w}_j^P - \widehat{w}_j^{P-1} \right] \tag{9.60}
$$

$$
-I_{ss}^P \gamma_j^2 \widehat{\phi}_j^P - I_{fs}^P \gamma_j^2 \frac{d\widehat{w}_j^P}{dx} + K_{sf}^P (\frac{d\widehat{w}_j^P}{dx} + \widehat{\phi}_j^P) - K_{fs}^P \frac{d^3 \widehat{w}^P}{d^3 x}
$$

$$
- K_{ss}^P \frac{d^2 \widehat{\phi}_j^P}{dx^2} = 0 \tag{9.61}
$$

Similarly, the force boundary conditions given by Equations 9.53 and 9.55 can be transformed as

$$
-K_{ff}^p \frac{d^3 w_j^p}{dx^3} + K_{sf}^p \left(\frac{dw_j^p}{dx} + \phi_j^p \right) - K_{fs}^p \frac{d^2 \phi_j^p}{x^2} = V_j \tag{9.62}
$$

$$K_{ff}^p \frac{d^2 w_j^p}{dx^2} + K_{fs}^p \frac{d\phi_j^p}{dx} = Q_j \tag{9.63}$$

$$K_{fs}^p \frac{d^2 w_j^p}{dx^2} + K_{ss}^p \frac{d\phi_j^p}{dx} = M_j$$

$$p = 1,\, 2, \ldots, P \text{ and } j = 0,\, 1, \ldots, n-1 \tag{9.64}$$

The spectral finite element has three degrees of freedom per node and per CNT wall, which are $\widehat{w}_j^p$, $\frac{\partial \widehat{w}_j^p}{\partial x}$, and $\widehat{\phi}_j^p$ for the pth CNT wall. The set of ODEs (Equations 9.56 to 9.61) for nano-composite beam with transverse and shear modes was obtained in a transformed wavelet domain. These equations are required to be solved for $\widehat{w}_j^p$, $\frac{\partial \widehat{w}_j^p}{\partial x}$, and $\widehat{\phi}_j^p$, and the actual solutions $w^p(x,t)$, $\frac{\partial w^p}{\partial x}(x,t)$, and $\phi^p(x,t)$ are obtained using inverse wavelet transform. The exact interpolating functions for an element of length L, obtained by solving Equations 9.56 and 9.57 to 9.61 respectively, are

$$\{\widehat{w}^1(x),\ \frac{\partial \widehat{w}^1(x)}{\partial x},\ \widehat{\phi}^1(x), \ldots, \widehat{w}^p(x),\ \frac{\partial \widehat{w}^p(x)}{\partial x},\ \widehat{\phi}^p(x), \ldots,$$

$$\widehat{w}^P(x),\ \frac{\partial \widehat{w}^P(x)}{\partial x},\ \widehat{\phi}^P(x)\}^T\ = [R][\Theta]\{a\} \tag{9.65}$$

where $[\Theta]$ is a $6P \times 6P$ diagonal matrix with diagonal terms $[e^{-k_1 x}, e^{-k_1(L-x)}, e^{-k_2 x}, e^{-k_2(L-x)}, e^{-k_3 x}, e^{-k_3(L-x)}, \ldots, e^{-k_{3P-2} x}, e^{-k_{3P-2}(L-x)}, e^{-k_{3P-1} x}, e^{-k_{3P-1}(L-x)}, e^{-k_{3P} x}, e^{-k_{3P}(L-x)}]$ and $[R]$ is a $2P \times 6P$ amplitude ratio matrix for each set of k_l, $l = 1$ to $3P$.

$$[\mathbf{R}] = \begin{bmatrix} R_{11} & \cdots & R_{16}, & \cdots\; \cdots & R_{1(6P-5)} & \cdots & R_{16P} \\ R_{21} & \cdots & R_{26}, & \cdots\; \cdots & R_{2(6P-5)} & \cdots & R_{26P} \\ \vdots & \cdots & \vdots, & \cdots\; \cdots & \vdots & \cdots & \vdots \\ R_{(P-1)1} & \cdots & R_{(P-1)6}, & \cdots\; \cdots & R_{(P-1)(6P-5)} & \cdots & R_{(P-1)6P} \\ R_{P1} & \cdots & R_{P6}, & \cdots\; \cdots & R_{P(6P-5)} & \cdots & R_{P6P} \end{bmatrix} \tag{9.66}$$

The wavenumbers k are obtained by substituting Equation 9.65 in Equations 9.56 and 9.57 to 9.61. This gives the characteristic equation. For example, for DWNT embedded composites, the characteristic equation is given as

$$\left[\begin{array}{cc} -\beta^2 I_0^1 + k^2(-\beta^2 I_{ff}^1 + K_{sf}^1) + k^4 K_{ff}^1 + C_1 & ik(\beta^2 I_{fs}^1 + K_{sf}^1) - ik^3 K_{fs}^1 \\ -ik(\beta^2 I_{fs}^1 + K_{sf}^1) + ik^3 K_{fs}^1 & (-\beta^2 I_{ss}^1 + K_{sf}^1) + k^2 K_{ss}^1 \\ 0 & -C_1 \\ 0 & 0 \end{array} \right.$$

$$\left. \begin{array}{cc} -C_1 & 0 \\ 0 & 0 \\ -\beta^2 I_0^2 + k^2(-\beta^2 I_{ff}^2 + K_{sf}^2) & ik(\beta^2 I_{fs}^2 + K_{sf}^2) - ik^3 K_{fs}^2 \\ \quad +k^4 K_{ff}^1 + C_1 & \\ -ik(\beta^2 I_{fs}^2 + K_{sf}^2) + ik^3 K_{fs}^1 & (-\beta^2 I_{ss}^2 + K_{sf}^2) + k^2 K_{ss}^2 \end{array} \right] [\mathbf{R}]\,\{a\} = 0$$

$$\tag{9.67}$$

The wavenumbers k are obtained by solving the above characteristic equation obtained by equating the determinant of the 4×4 matrix in Equation 9.67 to zero and the corresponding $[\mathbf{R}]$ is obtained using SVD of the matrix. The six different k obtained by solving Equation 9.67 correspond to the transverse, shear, and a purely imaginary higher order mode for each wall of the DWNT.

One characteristic of multiple beam model for MWNT embedded composite beam is that there are $P - 1$ cut-off frequencies, where the wavenumbers become zero, apart from the P shear modes, which have another P such frequencies.

Here, $\{a\}$ in Equation 9.65 are the unknown coefficients to be determined from transformed nodal displacements $\{\widehat{\mathbf{u}}^e\} = \{\widehat{w}_1^1 \ \frac{\partial \widehat{w}_1^1(x)}{\partial x} \ \widehat{\phi}_1^1 \ \widehat{w}_2^1 \ \frac{\partial \widehat{w}^1(x)}{\partial x} \ \widehat{\phi}_2^1 \ldots \widehat{w}_1^P$ $\frac{\partial \widehat{w}_1^P(x)}{\partial x} \ \widehat{\phi}_1^P \ \widehat{w}_2^P \ \frac{\partial \widehat{w}^P(x)}{\partial x} \ \widehat{\phi}_2^P \}$ and $\widehat{w}_1^p \equiv \widehat{w}^p(0)$, $\frac{\partial \widehat{w}_1^p}{\partial x} \equiv \frac{\partial \widehat{w}^p(0)}{\partial x}$, $\widehat{\phi}_1^p \equiv \widehat{\phi}^p(0)$ and similarly, $\widehat{w}_2^p \equiv \widehat{w}^p(L)$, $\frac{\partial \widehat{w}_2^p}{\partial x} \equiv \frac{\partial \widehat{w}^p(L)}{\partial x}$, $\widehat{\phi}_2^p \equiv \widehat{\phi}^p(L)$. Thus one can relate the nodal displacements and unknown coefficients as

$$\{\widehat{\mathbf{u}}^e\} = [\mathbf{T_1}]\{a\} \tag{9.68}$$

From the forced boundary conditions (Equations 9.62 to 9.64), nodal forces and unknown coefficients can be related as

$$\{\widehat{\mathbf{F}}^e\} = [\mathbf{T_2}]\{a\} \tag{9.69}$$

where $\{\widehat{\mathbf{F}}^e\} = \{\widehat{V}_1^1 \ \widehat{Q}_1^1 \ \widehat{M}_1^1 \ \widehat{V}_2^1 \ \widehat{Q}_2^1 \ \widehat{M}_2^1 \ldots \ldots \widehat{V}_1^P \ \widehat{Q}_1^P \ \widehat{M}_1^P \ \widehat{V}_2^P \ \widehat{Q}_2^P \ \widehat{M}_2^P\}$ and $\widehat{V}_1^p \equiv \widehat{V}^p(0)$, $\widehat{Q}_1^p \equiv \widehat{Q}^p(0)$, $\widehat{M}_1^p \equiv \widehat{M}^p(0)$ and $\widehat{V}_2^p \equiv \widehat{P}^p(L)$, $\widehat{Q}_2^p \equiv \widehat{Q}^p(L)$, $\widehat{M}_2^p \equiv \widehat{M}^p(L)$. From Equations 9.68 and 9.69 the dynamic stiffness matrix relating the transformed nodal forces and displacements can be obtained as

$$\{\widehat{\mathbf{F}}^e\} = [\mathbf{T_2}][\mathbf{T_1}]^{-1}\{\widehat{\mathbf{u}}^e\} = [\widehat{\mathbf{K}}^e]\{\widehat{\mathbf{u}}^e\} \tag{9.70}$$

where $[\widehat{K}^e]$ is the dynamic stiffness matrix.

9.4 Frequency domain analysis

Here, the WSFE model is used for frequency domain analysis of wave propagation in the nano-composite beam with embedded MWNT. As mentioned earlier, the beam is modeled as third order shear deformable using layer-wise theory and thus it has three modes for each nanotube wall. First, the spectrum and dispersion relations for double-walled carbon nanotube (DWNT)-aluminum matrix nano-composite beam are obtained assuming Euler-Bernoulli (EBT), Timoshenko (FSDT), and third order layer-wise shear deformation (TSDT) beam theories. Similar frequency dependent characteristics are studied in beams with three-walled carbon nanotube (TWNT) and different matrix materials, namely, polymer, metal, and ceramic, respectively.

All the numerical analysis in this and the next Sections are performed on a representative nano-composite beam with square cross-section of 6 nm × 6 nm. Different matrix materials including polymer, metal (aluminum), and ceramic (alumina) are considered whose material properties are taken as follows, for polymer : Young's modulus $E_p^m = 0.002$ TPa, shear modulus $G_p^m = 0.000769$ TPa, and density $\rho_p = 1000$ kg/m^3, for aluminum: $E_m^m = 0.07$ TPa, $G_p^m = 0.026$ TPa, and $\rho_m = 2700$ kg/m^3 and for ceramic: $E_c^m = 0.45$ TPa, $G_c^m = 0.19$ TPa, and $\rho_c = 2800$ kg/m^3. The inner diameter of MWNT is taken as 0.34 nm and the thickness of each wall is 0.34 nm. The elastic properties of MWNT are as follows, Young's modulus $E^c = 1.2$ TPa, shear modulus $G^c = 0.462$ TPa, and density $\rho^c = 1300$ kg/m^3. The van der Waals force interaction coefficient for the first wall becomes 0.0425 TPa as calculated from Equation 9.52.

In Figure 9.2, the non-dimensional wavenumbers (kh) are plotted for DWNT-aluminum matrix beam. The spectrum relations obtained by modeling the beam using EBT, FSDT, and TSDT are presented and compared in Figures 9.2(a), (b), and (c) respectively. It can be seen from Figure 9.2(a) that for a DWNT embedded aluminum matrix beam, EBT predicts two modes, one for each wall and one cut-off frequency (at 0.88 THz) as described in reference [11]. However, before the cut-off frequency, the wavenumber corresponding to the second mode has real as well as imaginary parts, which indicates that there is a propagating component of this mode. Due to the presence of the imaginary part, this wave will, however, attenuate while propagating. Thus, this wave is the so called inhomogeneous wave. Thus, there is non-zero speed even before the cut-off frequency. The non-dimensional wavenumbers plotted in Figure 9.2(b) for Timoshenko or first order shear deformable beam show that there exist two additional cut-off frequencies corresponding to the shear modes (0.40 and 7.9 THz for first and second shear modes respectively) of the two CNT walls. Unlike the flexural mode of the second CNT wall, these shear modes start propagating only after the cut-off frequencies. For TSDT, apart from the four shear and flexural propagating modes for DWNT embedded beam, there are two non-propagating modes corresponding to the higher order terms as shown in Figure 9.2(c). However, the present model using TSDT shows a considerable decrease in the shear cut-off frequencies while there is no significant change in cut-off frequency of the flexural mode for second CNT wall. The shear cut-off frequencies obtained from TSDT are at 0.30 THz and 4.45 THz. The corresponding dispersion relations or group speeds are presented in Figures 9.3(a), (b), and (c) for EBT, FSDT, and TSDT respectively. It can be seen from Figure 9.3 that as predicted by the wavenumber plots, for EBT model there are two propagating modes while for FSDT and TSDT, four such modes exist of which the shear modes propagate only after the corresponding cut-off frequencies.

Figures 9.4(a), (b) and 9.5(a), (b) show the spectrum relations for TWNT and TWNT embedded composites with polymer, aluminum, and ceramic as matrices respectively. These plots show the presence of six propagating modes

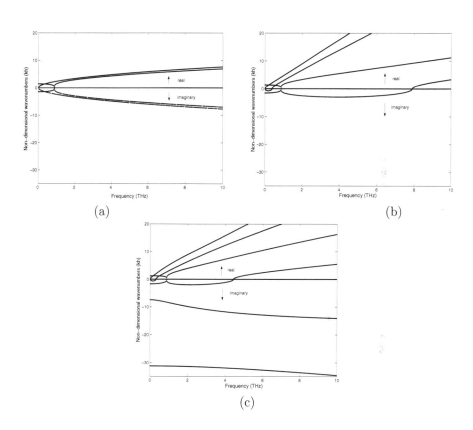

FIGURE 9.2: Spectrum relation for DWNT-aluminum matrix beam for (a) Euler-Bernoulli, (b) Timoshenko, and (c) third order layer-wise shear deformation models

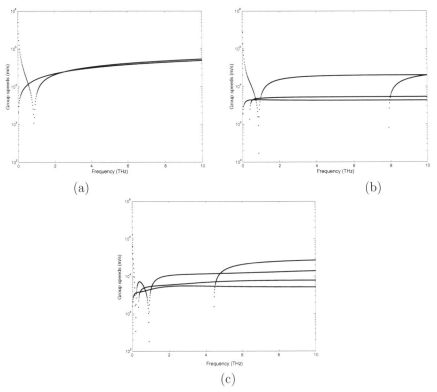

FIGURE 9.3:   Dispersion relation for DWNT-aluminum matrix beam for
(a) Euler-Bernoulli, (b) Timoshenko, and (c) third order layer-wise shear deformation models

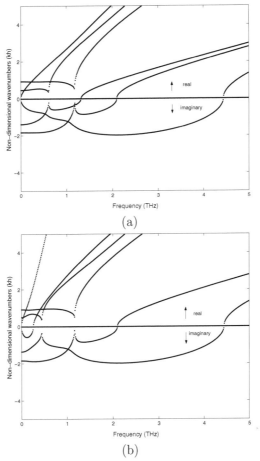

FIGURE 9.4: Spectrum relation for TWNT embedded beam for (a) no matrix and (b) polymer-matrix

with each CNT wall contributing a flexural and a shear mode. However, there are three more non-propagating modes corresponding to the higher order modes but are not shown in order to magnify the figure. It can be observed from the figures that by considering TWNT and TWNT embedded in three different matrix materials, considerable changes can be observed only in the first shear cut-off frequencies. These frequencies are 4.45, 0.25, 0.30, and 0.65 THz for TWNT and TWNT embedded polymer, aluminum, and ceramic matrices respectively. In Figures 9.6(a), (b) and 9.7(a), (b), the group speeds are plotted for TWNT and TWNT embedded composites with polymer, aluminum, and ceramic as matrices respectively. As expected, these dispersion relations show six propagating modes of which three shear modes, one for each CNT wall, propagate only after the respective cut-off frequencies.

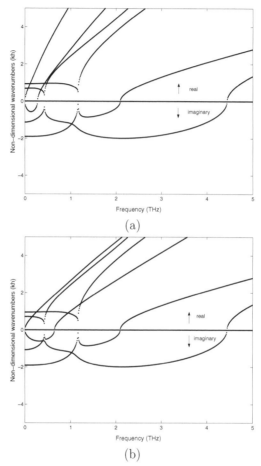

FIGURE 9.5: Spectrum relation for TWNT embedded beam for (a) aluminum-matrix and (b) ceramic-matrix

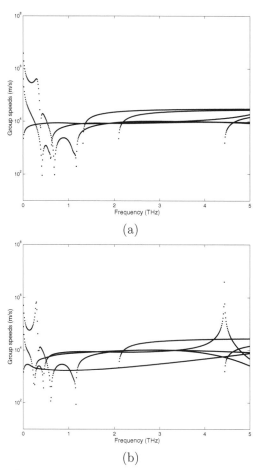

FIGURE 9.6: Dispersion relation for TWNT embedded beam for (a) no matrix and (b) polymer-matrix

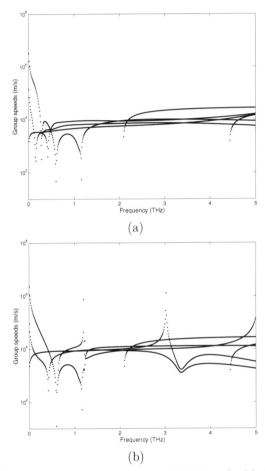

FIGURE 9.7: Dispersion relation for TWNT embedded beam for (a) aluminum-matrix and (b) ceramic-matrix

9.5   Time domain analysis

Next, the model developed is used to simulate time response in a finite length fixed-free beam due to broadband impulse load with frequency content at tera-Hertz level. The effects of the fraction of shear stress transfer between the CNT and matrix on the time response and also on the FRFs are investigated for polymer, metal, and ceramic matrix. Finally, the four coupled propagating modes in a DWNT aluminum matrix infinite beam are captured simultaneously using narrow banded modulated sinusoidal load. The material properties and geometry of the nano-composite beams are the same as that given in the previous section.

9.5.1   Response to broad-band impulse load

First the model is used for analysis of a finite length fixed-free beam due to unit broadband impulse load applied at the free end. The impulse load applied has a duration of 5.0 ps and frequency content of 1.0 THz. In Figure 9.8(a), the transverse velocities in DWNT-polymer composite beam of length 50 nm due to the impulse load applied at the tip in transverse direction are plotted. The time responses simulated by considering different fractions α as defined in Equation 9.13 of shear stress transfer between the outermost CNT and matrix are also compared for $\alpha = 0.1$, 0.5, and 1.0. It can be seen from the figure that effects of the above fractions α on the time response are negligible for such DWNT-polymer matrix. This is justified as the shear modulus ratio of CNT and polymer is very high causing the effect of α to be negligible. In Figure 9.8(b), the corresponding FRFs are plotted for different α. It can be seen that though the effect of α cannot be captured in the time response shown in Figure 9.8(a), the FRFs show certain changes at higher frequencies for $\alpha = 0.1$. Similar time and frequency responses are plotted in Figures 9.9 and 9.10, but for DWNT-aluminum and DWNT-ceramic composite beams respectively. It can be seen from Figures 9.9(a) and (b) that for DWNT-aluminum matrix composite, the effect of fraction of interfacial shear stress transfer is quite prominent, particularly for $\alpha = 0.1$, in both time responses and FRFs respectively. For DWNT-ceramic composite, the shear modulus ratio of CNT and matrix being much smaller than that for polymer and aluminum matrix, the effect of partial shear stress transfer at the interface is much more evident in both the time and frequency responses as can be seen from Figures 9.10(a) and (b).   From the above example, it is seen that the effect of polymer matrix on the response of CNT embedded nano-composite is negligible. Hence, while performing analysis, one can ignore the presence of polymer matrix and consider only the CNT for analysis. This aspect has large practical significance as most of carbon nanotube-based sensors which include bio-sensors are embedded on a polymer substrate.

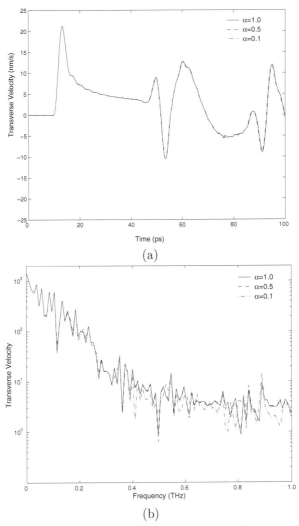

FIGURE 9.8:   Transverse tip velocities in DWNT-polymer composite fixed-free beam due to tip impulse load applied in transverse direction (a) time response and (b) FRF

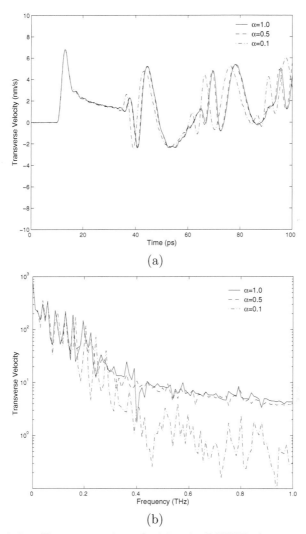

FIGURE 9.9: Transverse tip velocities in DWNT-aluminum composite fixed-free beam due to tip impulse load applied in transverse direction (a) time response and (b) FRF

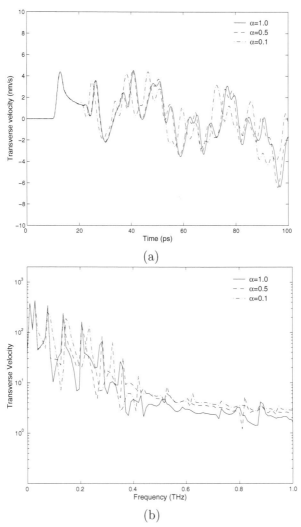

FIGURE 9.10: Transverse tip velocities in DWNT-ceramic composite fixed-free beam due to tip impulse load applied in transverse direction (a) time response and (b) FRF

9.5.2 Response to narrow-banded pulse

To study the presence of different propagating coupled modes in a MWNT embedded composite beam, the responses to narrow banded sinusoidal pulse modulated at a high frequency are simulated using the WSFE model. For such loading the waves propagate non-dispersively, that is, without changing the shape. The load is applied at a point on an infinite beam and the velocities are measured at another point at a distance L from the point of loading. Here, a DWNT-aluminum composite beam and $L = 50$ nm is considered. The load is modulated at 6.0 THz, such that the loading frequency is above the highest cut-off frequency which is approximately equal to 4.45 THz as shown in Figures 9.2(c) and 9.3(c). In Figure 9.11(a), the transverse velocity due to the pulse applied in transverse direction is plotted. The transverse velocity plotted is for the outermost wall and shows the presence of all four coupled modes, i.e., two shear and two flexural modes, one for each wall. The time of arrivals of the different propagating modes are those predicted from the dispersion relation given by Figure 9.3(c) and are approximately equal to 2.40, 4.70, 6.60, and 9.50 ps for shear and flexure modes of inner CNT wall and shear and flexure modes of outer CNT wall respectively. In Figure 9.11(b), similar response is plotted but using FSDT. As is seen from the dispersion plot in Figure 9.3(b), the cut-off frequency for the shear mode of the inner CNT wall is approximately 7.9 THz, which is higher than the loading frequency of 6.0 THz. As a result, Figure 9.11(b) shows only three propagating modes unlike Figure 9.11(a) for TSDT. Here also the time of arrival of the modes matches with that predicted by the corresponding dispersion plot. However, in Figures 9.11(a) and (b), the shear modes and the flexural modes corresponding to the other CNT wall are shown in a magnified scale.

9.6 Shell model of SWNT-polymer nano-composite

In this section, as mentioned earlier, the nano-composite comprised of SWNT and polymer matrix is considered as a continuum axisymmetric cylinder. The effects of the SWNT pull-out and the partial interfacial shear stress transfer are incorporated in the model as constraints which are imposed using the penalty matrix method [18]. The governing differential equations for such axisymmetric model and their reduction through Daubechies scaling functions are explained in Section 7.6. Similarly, the solution of these reduced ODEs using Bessel function and the WSFE formulation thereafter is also described in detail in Section 7.7. Here, the imposition of the constraints to incorporate the pull-out and partial shear stress transfer effects in the model is mostly described.

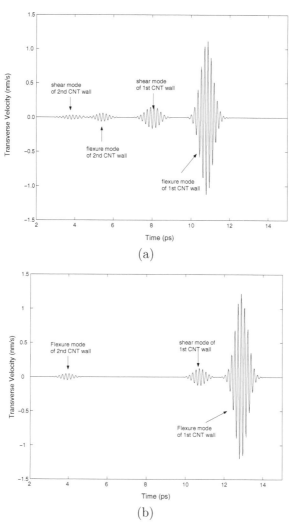

FIGURE 9.11:   Transverse velocities in a DWNT-aluminum composite infinite beam due to modulated sinusoidal pulse at 6.0 THz applied in transverse direction (a) third order layerwise deformation and (b) Timoshenko beam theories

9.6.1 Modeling of pull out and partial interfacial shear stress transfer

In Section 7.7, the formulation to obtain the elemental dynamic stiffness matrix $\widetilde{\mathbf{K}}^e$ has been explained. For the nano-composite structure consisting of a SWNT and polymer matrix, to obtain the global stiffness matrix, the elemental matrices, $\widetilde{\mathbf{K}}_c^e$ and $\widehat{\mathbf{K}}_m^e$, for the SWNT and the polymer respectively should be assembled. However, here the assembly is done using the penalty matrix method to impose constraints which account for the pull-out and the partial shear stress transfer.

The pull-out of the SWNT from the matrix is modeled as non-uniform axial strain at the interface, which can be written as

$$\frac{\partial u^m}{\partial z}\bigg|_{@\ r=r_o} = \vartheta \frac{\partial u^c}{\partial z}\bigg|_{@\ r=r_o} \tag{9.71}$$

where $u^m(r,z,t)$ and $u^c(r,z,t)$ are the axial displacement fields in the matrix and the SWNT respectively. r_o is the radius of the interface and ϑ is the factor defining amount of pull-out and is always less than one. After performing the Daubechies scaling function-based approximation as discussed in the earlier part of the section, Equation 9.71 can be reduced as

$$\widetilde{u}^m(r)\big|_{@\ r=r_o} = \vartheta \widetilde{u}^c(r)\big|_{@\ r=r_o} \tag{9.72}$$

Let $\{\widetilde{u}_1^m,\ \widetilde{w}_1^m,\ \widetilde{u}_2^m,\ \widetilde{w}_2^m\}$ be the nodal displacements at the two nodes at the outer and inner surfaces of the matrix respectively. Similarly, let $\{\widetilde{u}_1^c,\ \widetilde{w}_1^c,\ \widetilde{u}_2^c,\ \widetilde{w}_2^c\}$ be the modal displacements at the outer and the inner surfaces of the SWNT respectively. Thus, Equation 9.72 can be written as

$$\widetilde{u}_2^m = \vartheta \widetilde{u}_2^c \tag{9.73}$$

The above equation can be written in matrix form as

$$\begin{bmatrix} 0\ 0\ 1\ 0\ -\vartheta\ 0\ 0\ 0 \end{bmatrix} \{\widetilde{\mathbf{u}}\} = 0 \quad \text{or,} \quad [\mathbf{C_1}]\{\widetilde{\mathbf{u}}\} = 0 \tag{9.74}$$

where $\{\widetilde{\mathbf{u}}\}$ is the total nodal displacement vector $\{\widetilde{u}_1^m,\ \widetilde{w}_1^m,\ \widetilde{u}_2^m,\ \widetilde{w}_2^m,\ \widetilde{u}_1^c,\ \widetilde{w}_1^c,\ \widetilde{u}_2^c,\ \widetilde{w}_2^c\}$. Next, apart from the constraint given by Equation 9.71, the radial displacements $w^m(r,z,t)$ and $w^c(r,z,t)$ should be uniform at the interface $r = r_o$,

$$w^m\big|_{@\ r=r_o} = w^c\big|_{@\ r=r_o} \tag{9.75}$$

Similar to Equation 9.72, the transformed form of Equation 9.75 can be written as

$$\widetilde{w}^m(r)\big|_{@\ r=r_o} = \widetilde{w}^c(r)\big|_{@\ r=r_o} \quad \text{or,} \quad \widetilde{w}_2^m = \widetilde{w}_2^c \tag{9.76}$$

This again can be written in a matrix form as

$$\begin{bmatrix} 0\ 0\ 0\ 1\ 0\ -1\ 0\ 0 \end{bmatrix} \{\widetilde{\mathbf{u}}\} = 0 \quad \text{or,} \quad [\mathbf{C_2}]\{\widetilde{\mathbf{u}}\} = 0 \tag{9.77}$$

Thus, for the pull-out modeling, the total constraint equation can be written as follows,

$$[\mathbf{C_p}]\{\widetilde{\mathbf{u}}\} = 0 \quad \text{where} \quad [\mathbf{C_p}] = \begin{bmatrix} \mathbf{C_1} \\ \mathbf{C_2} \end{bmatrix} \tag{9.78}$$

The final constraint Equation 9.78 can be imposed on the WSFE model using the penalty matrix method [18], which is given as

$$\left([\widetilde{\mathbf{K}}] + [\mathbf{C_p}]^T[\alpha][\mathbf{C_p}]\right)\{\widetilde{\mathbf{u}}\} = \{\widetilde{F}\} \tag{9.79}$$

where $[\mathbf{C_p}]^T[\alpha][\mathbf{C_p}]$ is the penalty matrix and $[\alpha]$ is a diagonal matrix with diagonal terms called penalty numbers. The penalty matrix is usually singular. The accurate imposition of constraint requires the value of the parameter α_i to be very large. $\{\widetilde{F}\}$ is the total nodal force vector, $\{\widetilde{\sigma}_{z1}^m, \widetilde{\sigma}_{r1}^m, \widetilde{\sigma}_{z2}^m, \widetilde{\sigma}_{r2}^m, \widetilde{\sigma}_{z1}^c, \widetilde{\sigma}_{r1}^c, \widetilde{\sigma}_{z2}^c, \widetilde{\sigma}_{r2}^c\}$ and $[\widetilde{\mathbf{K}}]$ is given as

$$[\widetilde{\mathbf{K}}] = \begin{bmatrix} \widetilde{\mathbf{K}}_\mathbf{m}^\mathbf{e} & 0 \\ 0 & \widetilde{\mathbf{K}}_\mathbf{c}^\mathbf{e} \end{bmatrix} \tag{9.80}$$

As $[\alpha]$ grows, the constraints are more nearly satisfied [18]. However, much higher values of $[\alpha]$ may cause ill-conditioning of the modified matrix.

Similarly, the constraint for the partial interfacial shear stress transfer can be written as

$$\tau_{rz}^c\Big|_{@ \ r=r_o} = \varsigma\tau_{rz}^m\Big|_{@ \ r=r_o} \tag{9.81}$$

where τ_{rz}^c and τ_{rz}^m are the shear stress in the SWNT and the matrix respectively. ς is the fraction of the shear stress transfer between the matrix and the SWNT, and has a value less than one. The above constraint given by Equation 9.81 can be represented in a matrix form similar to Equation 9.74 as

$$[\mathbf{C_3}]\{\widetilde{\mathbf{u}}\} = 0 \tag{9.82}$$

$$[\mathbf{C_3}] = \frac{1}{2\pi r_o L_z}\left[\varsigma\widetilde{\mathbf{K}}_\mathbf{m}^{\mathbf{e31}} \ \varsigma\widetilde{\mathbf{K}}_\mathbf{m}^{\mathbf{e32}} \ \varsigma\widetilde{\mathbf{K}}_\mathbf{m}^{\mathbf{e33}} \ \varsigma\widetilde{\mathbf{K}}_\mathbf{m}^{\mathbf{e34}} \ -\widetilde{\mathbf{K}}_\mathbf{c}^{\mathbf{e11}} \ -\widetilde{\mathbf{K}}_\mathbf{c}^{\mathbf{e12}} \ -\widetilde{\mathbf{K}}_\mathbf{c}^{\mathbf{e13}} \ -\widetilde{\mathbf{K}}_\mathbf{c}^{\mathbf{e14}}\right]$$

where $\widetilde{\mathbf{K}}_{\mathbf{c/m}}^{\mathbf{eij}}$ represent the element with indices i and j of the matrices $\widetilde{\mathbf{K}}_{\mathbf{c/m}}^\mathbf{e}$. In addition to the above Equation 9.82, the constraint of uniform radial displacement at the interface given by Equation 9.77 also applies here. Thus, for modeling partial interfacial shear stress transfer, the total constraint equation can be written similar to Equation 9.78 as

$$[\mathbf{C_s}]\{\widetilde{\mathbf{u}}\} = 0 \quad \text{where} \quad [\mathbf{C_s}] = \begin{bmatrix} \mathbf{C_3} \\ \mathbf{C_2} \end{bmatrix} \tag{9.83}$$

The matrix $[\mathbf{C_s}]$ can be used in Equation 9.79 in place of $[\mathbf{C_p}]$ to obtain the modified dynamic stiffness matrix incorporating the effect of the partial shear stress transfer.

9.7   Time domain analysis

In this section, numerical experiments are performed to study the effects of the pull-out of the SWNT from the matrix and the partial interfacial shear stress transfer on the wave responses. Though the pull-out effects are predominant on the axial wave propagation, the radial wave propagation also gets modified due to the radial-axial coupling existing in such finite length axisymmetric structures.

The example considered here consists of a $(10, 10)$ SWNT with radius $R = 0.678$ nm. The radius R of a (N, N) SWNT is calculated as $R = 3Na_{C-C}/(2\pi)$ [14], where a_{C-C} is carbon bond length and equal to 0.142 nm. The thickness $2h$ of the SWNT is considered as 0.34 nm. The inner and outer radii of the SWNT are $r_i = R - h$ and $r_o = R + h$ respectively. The thickness of the surrounding matrix is $2h_m = 2$ nm and, thus, the outer radius of the nano-composite cylinder is $r_m = r_o + 2h_m$. The length of the structure is taken as $L_z = 25$ nm with free-free ends. The bulk material properties of the SWNT are Young's modulus $Eh = 360$ J/m^2 [102], mass density $\rho h = 2270 \times 0.34$ kg/m^3, and Poisson's ratio $\nu = 0.2$ [98]. In particular, these parameters are not dependent on the definition of thickness $2h$ [102]. Similarly, the material properties of the polymer are as follows, Young's modulus $E = 15$ GPa, mass density $\rho = 1500$ kg/m^3, and Poisson's ratio $\nu = 0.3$.

The applied load is a broadband impulse loading with duration 0.1 ps and a frequency content 45 THz. The load is applied along the outer surface of the nano-composite in the axial or the radial direction. The spatial distribution of the impulse load is given as $F(z) = e^{-((Z-k)/\varpi)^2}$, where k and ϖ are constants which determine the spatial location and distribution of the load. Here, the load is applied at the free end and has a very small spatial width to simulate point load.

9.7.1   Pull-out of SWNT

In Figures 9.12(a) and (b), the axial velocities of the SWNT and the polymer at the interface with $r = r_o$ are plotted for different values of ϑ given in Equation 9.71. The axial velocities are measured at the free end where the load is applied. The values of ϑ considered are 0.1, 0.5, and 0.9 and $\vartheta = 1.0$. The last value indicates perfect bonding between the polymer matrix and the SWNT. Thus, for $\vartheta = 1.0$, the wave responses of the SWNT and the polymer are similar. It can be observed from Figure 9.12(a), that as the difference in the axial strain between the SWNT and the polymer increases, the velocity of the SWNT increases particularly at the later part of the response. In other words, the axial velocity of the SWNT increases much for $\vartheta = 0.1$ compared to that for perfect bonding or no pull-out of the SWNT from the matrix. However, the increase in the velocity can be observed in the later part of the

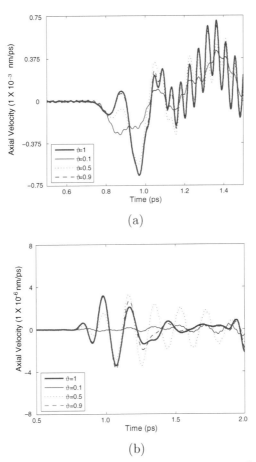

FIGURE 9.12: Axial velocity at the interface for the (a) SWNT and (b) polymer matrix

response between 0.15 to 0.2 ps. Similarly, for $\vartheta = 0.5$ also, a substantial increase in the axial velocity is observed, while for $\vartheta = 0.9$ the velocity is almost similar to that for no pull-out condition. On the other hand, for the polymer, the axial velocity for $\vartheta = 0.1$ is considerably less than the velocity in a perfectly bonded SWNT.

Figures 9.13(a) and (b) show the frequency response functions (FRFs) of the axial velocities of the SWNT shown in Figure 9.12(a) for frequency ranges $0-11$ and $30-38$ THz respectively. The natural frequencies with considerable amplitude lie in these ranges. From Figure 9.13(a), for the frequency range $0-11$ THz, it can be seen that there is not much shift in the natural frequencies. However, for $\vartheta = 0.1$, the amplitude is less than that for the no pull-out condition, except at the frequency ≈ 5.5 THz, where the amplitude is more.

Similarly, for $\vartheta = 0.5$, there is a considerable increase in the amplitude for the frequency ≈ 5.5 THz. This observation is an explanation of nearly single frequency oscillatory motion in Figure 9.12(a) for $\vartheta = 0.1$ and 0.5. This observation is in tune with that reported in the reference [33], where the response of the SWNT after pull-out from the polymer matrix is found to be oscillatory. In the frequency range $30-38$ THz, shown in Figure 9.13(b), there are some shifts in the natural frequencies. However, the amplitude of these frequencies are very small compared to the frequencies in the range $0-11$ THz, shown in Figure 9.13(a). The FRFs for the axial velocities in the polymer at the interface shown in Figure 9.12(b) are plotted in Figures 9.14(a) and (b) for frequency ranges $0-11$ and $30-38$ THz respectively. It can be seen from the Figure 9.14(a), that pull-out of the SWNT does not cause any shift in the natural frequencies and even there is no increase in the amplitudes. However, in Figure 9.14(b) some shift in the natural frequencies are observed. However, the amplitudes of these frequencies are small compared to the frequencies in the range $0-11$ THz shown in Figure 9.14(b) and thus do not contribute to the wave responses.

Figures 9.15(a) and (b) show the shear stress at the interface for the SWNT and the polymer matrix respectively. The shear stresses are plotted for different values $\vartheta = 0.1$, 0.5, 0.9 and 1.0 denotes perfect bonding. It can be seen that the interfacial shear stresses are affected considerably by the pull-out of the SWNT. It should be mentioned here that even for perfect bonding between the SWNT and the matrix, there is a discontinuity in the shear stress at the interface as shown by the Figures 9.15(a) and (b).

In Figure 9.16, the radial velocity at the interface is plotted for different values of $\vartheta = 0.1$, 0.5, 0.9, and 1.0. The radial wave propagation results due to the impulse load as described earlier. However, here the load is applied in the radial direction along the outermost surface of the nano-composite instead of the axial direction as in the previous examples. The effects of the SWNT pull-out are not very prominent for the radial wave propagation and the difference in the responses are due to the axial-radial coupling present in such finite length axisymmetric structures. This coupling in finite length SWNT has been reported in literature on experimental studies and atomistic simulations [66].

9.7.2 Partial interfacial shear stress transfer

Here, axial wave propagation in the nano-composite structure is studied with enforced partial shear stress transfer at the interface as given in Equation 9.81. Here, the partial stress transfer has been accounted for through a kinematic constraint as explained in the last section. In Figure 9.17(a) and (b), the axial velocities at the interface are plotted respectively for the SWNT and the polymer for different fractions of stress transfer ς given by Equation 9.81. It can be observed that the response of the SWNT as shown in Figure 9.17(a) is not affected considerably by the partial stress transfer.

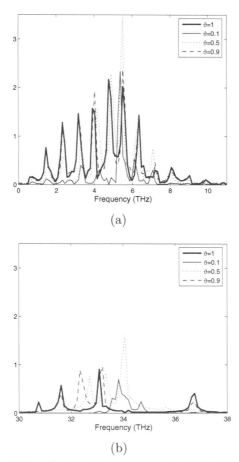

FIGURE 9.13: FRFs of axial velocity at the interface for the SWNT between frequency ranges (a) 0–11 THz and (b) 30–38 THz

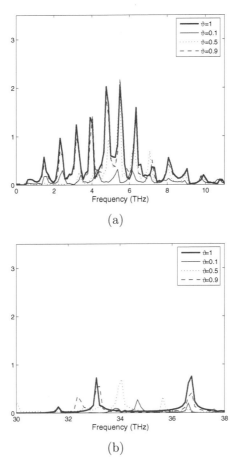

(a)

(b)

FIGURE 9.14: FRFs of axial velocity at the interface for the polymer between frequency ranges (a) 0–11 THz and (b) 30–38 THz

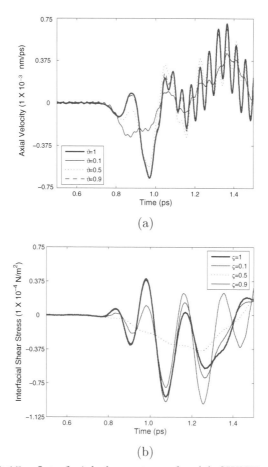

(a)

(b)

FIGURE 9.15: Interfacial shear stress for (a) SWNT and (b) polymer matrix

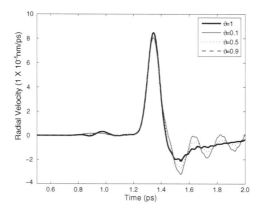

FIGURE 9.16: Radial wave velocity at the interface

This is in congruence with that obtained in [64]. In contrast, the axial velocity of the matrix at the interface shown in Figure 9.17(b) is considerably affected by the partial stress transfer, particularly for $\varsigma = 0.1$. The effect of partial stress transfer is not considerable on the response of the SWNT as the shear modulus G^c of the SWNT is nearly the same as the effective shear modulus, G, of the nano-composite. As a result, the factor ς lying between 0 to 1.0 does not have much effect on the response of the SWNT. On the other hand, the shear modulus, G^m, of the polymer matrix is very small compared to the effective shear modulus G of the nano-composite. Thus, the response of the polymer is largely affected by the variation of ς. However, the effect of partial stress transfer on the matrix is more important compared to the effect on the SWNT. Here, a RVE is considered and the dimension of the polymer matrix is comparable to the dimension of the SWNT. However, the present study is done with the motivation to understand these effects on nano-composites where there is a scale difference between the dimensions of the nano-composite and the SWNT, which acts as reinforcing materials. In such cases, the effects on the overall nano-composites, which is mostly matrix material with 0.1 % to 1.0 % CNTs, is only important.

To summarize it can be said that the present chapter is an extension of Chapter 8, to analyze the different dynamic behavior of nano-composites to explore their potential as a functional, lightweight, high stiffness material. Even here, the nano-composites are modeled using continuum theories. These models, however, consider imperfect bonding between the CNT and surrounding matrix to simulate phenomena like pull-out of CNT from the matrix and partial stress transfer. Most importantly, WSFE is found very effective in these analyses and helped in simulating several non-classical dynamical behaviors of nano-composites not explored before in literature.

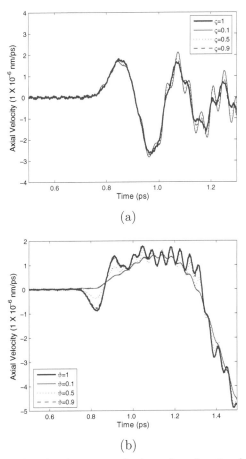

FIGURE 9.17:   Axial velocity at the interface for the (a) SWNT and (b) polymer matrix for partial interfacial shear stress transfer

Chapter 10

Inverse Problems

As mentioned earlier, wave propagation analysis finds important applications in inverse problems of force reconstruction, identification of material properties, damage detection, and health monitoring. In this chapter we limit ourselves to two important applications of identification of the source from the measured response and the detection of the damages using suitable models and measured responses.

10.1 Force reconstruction

One of the biggest advantages of the spectral approach is in the performance of inverse problems. This is because, in spectral approach, system transfer function, which is one of the principle ingredients to perform inverse problem, is obtained as a direct byproduct. Force identification is one such inverse problem. Identification of dynamic force from the experimentally measured response at some point is a problem of wide applicability. The convenience of using Fourier transform-based spectral element to predict the force history from the measured responses has been demonstrated for mono-material beam [24], bi-material [27] beam, isotropic [25], and orthotropic [26] plates. Similar force identifications were presented for isotropic layered media [85] and for inhomogeneous layered media [9, 10]. Two problems that are associated with force identification are the structures of short lengths and the measured incomplete data. These two problems are interrelated. Short lengths cause multiple reflections which make the response not die down within the chosen time window. These are necessary for accurate prediction of force in Fourier transform-based methods. Filtering these would make the data incomplete, which may predict spurious forces. Normally experimentally recorded signals are bound to be truncated. The truncation point proves to be an important factor to determine the accuracy of the reconstructed force. This problem is even more severe in a dispersive system, where the wave response will not die down completely within the chosen time window. All these problems occur in Fourier transform-based methods due to assumed periodicity in both forward and inverse Fourier transforms. This problem is expected to be solved using wavelet transform as no such periodicity is assumed in the formulation of

WSFE. In addition, the localized nature of the wavelet basis functions enables easy handling of finite geometries. Hence, as explained in the case of forward problem, the accuracy of the identified force using WSFE is independent of the point of truncation.

Wavelet-based techniques have been used for inverse problems of measuring temperature from the sideways heat equation [82, 83, 31]. Meyer and Daubechies wavelets were used to approximate the time derivative and the results were compared with Fourier-based approximation. In reference [29], reconstruction of force is done from experimentally recorded wave response of an impacted plate with a hole using wavelet representation of unknown load and FE method. In the present work, truncated FE responses obtained at certain points are used as surrogate experimental responses. These truncated FE responses are given as input to spectral element solvers and the force data are reconstructed by performing inverse analysis. Though these truncated responses obtained from FE simulations are used as input, they are free from several complexities associated with experimentally measured responses. Reference [96] presents a continuous wavelet transform-based method for identification of excited modes, moment of rupture, and de-noising in an impact test.

The basic idea of spectral finite element is to obtain a system transfer function that relates the input to the output. For example, in the WSFE formulation for elementary rod given in Section 5.1 of Chapter 5, the nodal displacements are obtained in terms of the nodal forces through elemental dynamic stiffness matrix, given by Equation 5.35. Again, using Equation 5.36, the constants $\{a\}$ can be derived and these give the displacement $\widehat{u}(x)$ at any spatial location x from Equation 5.26. These finally relate the transformed nodal forces with the transformed displacement at any arbitrary spatial position. It can be said that the transform of the load $\widehat{F}$ and displacements $\widehat{u}$ are related through transfer function $\widehat{G}(x)$ as

$$\widehat{u}(x) = \widehat{G}(x)\widehat{F}(x) \qquad (10.1)$$

Other parameters like stress, velocity, strain, etc. are related to load and displacements and can be used as input or output with required modification in $\widehat{G}(x)$. The inverse problem of calculating the input force $\widehat{F}$ can be done as

$$\widehat{F}(x) = \widehat{u}(x)/\widehat{G}(x) \qquad (10.2)$$

The system transfer function $\widehat{G}(x)$ is obtained as follows. After the global $[\widehat{K}]$ is formed, the structure is solved for unit impulse at the desired location. This will directly give $\widehat{G}(x)$. The measured response $u(x,t)$ in the time domain is then transformed to the wavelet domain to get $\widehat{u}(x)$. Then Equation 10.2 is used to obtain the force history in wavelet domain, which is then transformed by inverse wavelet transform to obtain required force history.

10.2 Numerical examples of impulse force reconstruction

The earlier examples presented in Section 5.4 of Chapter 5 demonstrate the presence of wraparound problems in Fourier transform-based solutions. When such solutions are used to perform inverse problems such as force identification on a shorter waveguide, due to time window limitation, the solution obtained may yield a highly distorted force history. This is where the formulated WSFE will be of great utility. Its use in force identification is demonstrated in here.

The example used first is an aluminum rod with Young's modulus $E = 70$ GPa and density $\rho = 2.7 \times 10^3$ kg/m^3. The rod is fixed at one end and an axial impulse load is applied at the free end. First, force identification from measured axial velocity is done considering undamped condition. The longitudinal velocity at the mid point of the rod due to tip axial load obtained using FE is used as input. The unit impulse load described in Section 5.4 and shown in Figure 5.2 is used as the applied load to obtain the FE solution. The rod is modeled with 400, 3 noded plane stress triangular elements and the response is presented in Figure 10.1. Figure 10.2(a) shows the force reconstructed from the above response truncated at $T_c = 512~\mu$s, which shows an excellent match with applied force. The forces identified from responses truncated at different points are the same irrespective of the point of truncation (T_c). Thus this spectral element can efficiently reconstruct the force even from responses recorded for a small time duration. Similar to the forward problem, Fourier transform-based techniques cannot be used for performing inverse problems in undamped finite length structure.

Though the above example shows the advantage of WSFE in force identification, the reconstruction will involve certain problems when experimental data are used. This is because the experimental responses are never free from noise which causes distortion in the reconstructed force. In Figure 10.4(a), a noisy response is simulated by adding white noise to the response shown in Figure 10.1, to produce a signal to noise ratio of 7.5. The reconstructed force is shown in Figure 10.4(b) and it can be seen that it is highly distorted. This example shows that as mentioned earlier, when experimentally measured responses are used as input, de-noising of the signal is required prior to reconstruction.

Next, we deal with damped structures to compare the results obtained using WSFE and the corresponding Fourier transform-based method. The response used as input is similar to that shown in Figure 10.1 except that a damping of $\eta = 1.0$ is considered. The force reconstructed using wavelets in this case is very much similar to that presented in Figure 10.2(a). Figure 10.2(b) presents the forces identified using Fourier transform-based method for different truncation points (T_c). It can be seen that for $T_c = 512~\mu$s, the force obtained is highly inaccurate and gradually gets refined with increased T_c.

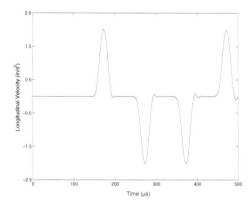

FIGURE 10.1: Longitudinal velocity in rod measured at mid-point ($x = 10$ in) due axial impact load at tip ($x = 20$ in) using two-dimensional FE

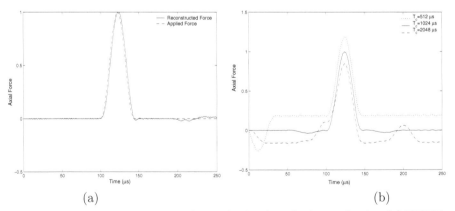

FIGURE 10.2: Reconstructed impulse load applied to rod using (a) WSFE and (b) Fourier transform-based technique

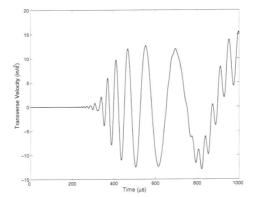

FIGURE 10.3:   Transverse velocity in beam measured at mid-point ($x = 5$ in) due transverse impact load at tip ($x = 10$ in) using 2-D FE

Similar experiments are performed for force identification from recorded transverse velocities in aluminum beams. The cross-sectional dimensions of the beam are same as the rod except that the length considered is $L = 10$ in. The flexural velocity measured at the mid-point of the beam due to tip impact load simulated using FE and shown in Figure 10.3 is used as input. The impulse loading and the 2-D FE mesh used are same as that used for the rod. Figure 10.5(a) shows the force reconstructed from the above response truncated at $T_c = 512$ μs using WSFE. As stated for the rod, force identification using WSFE is independent of the point of truncation (T_c).

Finally force identification is done for the two-dimensional frame structure shown in Figure 5.7 from the flexural response obtained through FE analysis and is shown in Figure 5.8. The impulse loading is same as that used for rod and beam. The FE meshing is done with 5000 one-dimensional beam elements for each of the three members of the frame and a total of 15000 elements. Similar to the previous force identification experiments done, for two-dimensional frames WSFE reconstructs the impulse force accurately, irrespective of T_c and the plot is presented in Figure 10.6(a). For this problem the forces identified using Fourier transform-based methods and considering a damping of $\eta = 1.0$ are plotted in Figure 10.6(b). It can be seen that for accurate reconstruction of force the method requires the truncation point $T_c = 4096$ μs which is higher than that required in previous examples of beam and rod. This can be justified as the response in this problem contains multiple reflections. However, these problems prove the efficiency of WSFE in force reconstruction even for such complex structures.

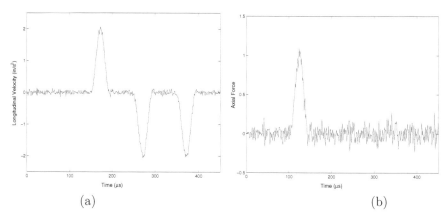

(a)   (b)

FIGURE 10.4:   Reconstruction of impulse load applied to rod (a) response used as input (Figure 10.1 with simulated white noise) and (b) reconstructed impulse load

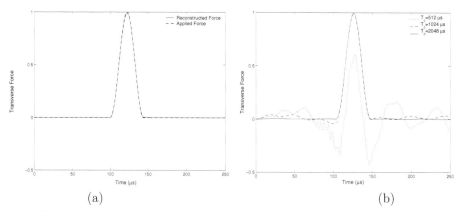

(a)   (b)

FIGURE 10.5:   Reconstructed impulse load applied to beam (a) WSFE and (b) Fourier transform-based technique

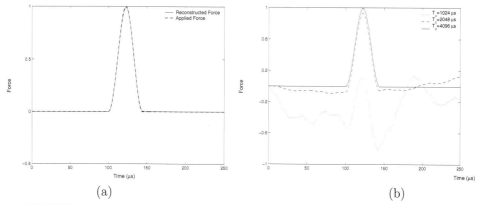

FIGURE 10.6: Reconstructed impulse load applied to 2-D frame (a) WSFE and (b) Fourier transform-based technique

10.3   Damage modeling and detection

Another inverse problem of great practical value is the detection of damages in structures. As mentioned earlier, the diagnostic waves are extensively used for this purpose. Wave propagation problems deal with high frequency excitations and thus help to identify the presence of very small damages. These wave-based techniques can thus be used to detect the minute defects which occur at the onset of the damage and then propagate causing failure of the structure. Proper use of these techniques first requires good knowledge of the effects of damages on the wave characteristics. This needs accurate and computationally efficient modeling of the damaged structures. In this chapter, two such problems of damage modeling and their detection are analyzed. In either case, WSFE technique is used to simulate the wave propagation in such damaged structures and study the effects of damages on the wave behavior.

Here, the problem deals with modeling and detection of de-lamination in composite beams based on WSFE formulation. Though composites have several favorable properties, which are encouraging their use in different parts of an aircraft, the behavior of composites under damage is very complicated and not well understood due to anisotropic nature of the material. The transverse tensile and inter-laminar shear strengths of composites are much lower than the in-plane properties. This makes such structures very much prone to defects like matrix cracking, fiber fracture, fiber de-bonding, de-lamination/inter-laminar de-bonding, of which de-lamination is most common, and can grow, thus reducing the life of the structure.

In this chapter, the WSFE model of a composite beam with embedded de-lamination is formulated in a similar way as presented by [71]. The technique used to model a through width de-lamination subdivides the beam into base-laminates and sub-laminates along the line of de-lamination. The base-laminates and sub-laminates are treated as structural waveguides, and kinematics are enforced along the connecting line(s). These waveguides are modeled as Timoshenko beams with elastic and inertial coupling and the corresponding spectral elements have three degrees of freedom, namely, axial, transverse, and shear displacements at each node. The internal spectral elements in the region of de-lamination are assembled assuming constant cross-sectional rotation and equilibrium at the interfaces between the base-laminates and sub-laminates. Finally, the redundant internal spectral element nodes are condensed out to form two noded spectral elements with embedded de-lamination.

Recently, wavelet analysis of wave propagation for damage detection is widely reported in literature [52, 91, 106]. In the present work, Daubechies wavelet-based post processing of measured responses of damaged beam is done for simultaneous de-noising and arrival time-based damage detection. Here, the simulated responses of a de-laminated beam modeled using WSFE are used as substitutes for experimental results. Further, artificial white noise is added to the simulated response. The main aim of wavelet analysis here is to filter out the unnecessary frequency components from the response for better extraction of arrival time which otherwise cannot be obtained directly from the response. De-noising is done by universal soft thresholding at the wavelet levels mainly corresponding to higher frequencies [23].

10.4   Modeling of de-lamination in composite beam

The first step in modeling of a composite beam with embedded de-lamination is to develop a WSFE for the undamaged composite waveguide. Here, the composite beam is modeled as a Timoshenko beam with axial, transverse, and shear degrees of freedom. Similarly, the WSFE model has been formulated and validated in Chapter 5, except that an additional contractional degree of freedom is considered. Thus, here, the formulation of the WSFE for the undamaged composite beam is not repeated. However, in this section the modeling of de-lamination is presented.

As said earlier, here the modeling of embedded de-lamination is done according to the method presented in reference [71]. Figure 10.7(a) shows the de-laminated beam broken down into two base-laminates, i.e., laminates 1, 2 on either sides of the de-lamination and two sub-laminates, i.e., laminates 3, 4 on its bottom and top respectively. Each of these laminates are considered as individual waveguides and modeled as coupled composite Timoshenko beams using WSFE described in the last sub-section. These internal spectral elements with the nodes are shown in Figure 10.7(b) and are assembled as-

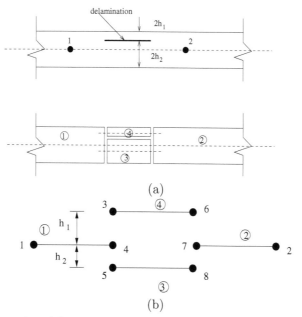

FIGURE 10.7: (a) Modeling of an embedded de-lamination with base-laminates and sub-laminates. Waveguides 1, 2: base laminates and 3, 4: sub-laminates; (b) representation of the base-laminates and sub-laminates by spectral elements

suming constant cross-sectional slope/rotation and force equilibrium at their interfaces. The first assumption of constant and continuous slope at the two interfaces between the sub-laminates and base-laminates gives the following kinematic relation

$$\widehat{\mathbf{u}}_3 = \mathbf{S}_1\widehat{\mathbf{u}}_4 \quad \widehat{\mathbf{u}}_5 = \mathbf{S}_2\widehat{\mathbf{u}}_4 \quad \widehat{\mathbf{u}}_6 = \mathbf{S}_1\widehat{\mathbf{u}}_7 \quad \widehat{\mathbf{u}}_8 = \mathbf{S}_2\widehat{\mathbf{u}}_7 \qquad (10.3)$$

where $\widehat{\mathbf{u}}_i$ represents the nodal displacement vector of the ith node in Figure 10.7(b). Thus all the nodal displacements of the sub-laminate elements 3 and 4 are represented in terms of nodal displacements of base-laminate elements 1 and 2. $\mathbf{S}_1$ and $\mathbf{S}_2$ are the 3×3 transformation matrices in terms of top and bottom sub-laminate thicknesses $2h_1$ and $2h_2$, shown in Figure 10.7(a) and are given as

$$\mathbf{S}_1 = \begin{bmatrix} 1 & 0 & h_1 \\ 0 & 1 & 0 \\ 0 & 0 & 0 \end{bmatrix} \qquad \mathbf{S}_2 = \begin{bmatrix} 1 & 0 & -h_2 \\ 0 & 1 & 0 \\ 0 & 0 & 0 \end{bmatrix} \qquad (10.4)$$

Next, using force equilibrium at the two interfaces of sub- and base-laminates, the following relations can be derived

$$\widehat{\mathbf{f}}_4 + \mathbf{S}_1^T\widehat{\mathbf{f}}_3 + \mathbf{S}_2^T\widehat{\mathbf{f}}_5 = 0 \quad \widehat{\mathbf{f}}_7 + \mathbf{S}_1^T\widehat{\mathbf{f}}_6 + \mathbf{S}_2^T\widehat{\mathbf{f}}_8 = 0 \qquad (10.5)$$

where $\widehat{\mathbf{f}}_i$ represents the nodal displacement vector of the ith node, shown in Figure 10.7(b). Let $\widehat{\mathbf{K}}^j$ be the dynamic stiffness matrix for the jth element with p and q nodes rewritten using 3×3 sub-matrices as

$$\begin{bmatrix} \widehat{\mathbf{K}}_{11}^j & \widehat{\mathbf{K}}_{12}^j \\ \widehat{\mathbf{K}}_{21}^j & \widehat{\mathbf{K}}_{22}^j \end{bmatrix} \begin{Bmatrix} \widehat{\mathbf{u}}_p \\ \widehat{\mathbf{u}}_q \end{Bmatrix} = \begin{Bmatrix} \widehat{\mathbf{f}}_p \\ \widehat{\mathbf{f}}_q \end{Bmatrix} \qquad (10.6)$$

Using the transformation equations, Equations 10.3 and 10.6, the four internal spectral elements are assembled. Further, using Equation 10.4, we get

$$\begin{bmatrix} \widehat{K}_{11}^1 & \widehat{K}_{12}^1 & 0 & 0 \\ \widehat{K}_{21}^1 & \widehat{K}_{22}^1 + S_1^T\widehat{K}_{11}^4 S_1 + S_2^T\widehat{K}_{11}^3 S_2 & S_1^T\widehat{K}_{12}^4 S_1 + S_2^T\widehat{K}_{12}^3 S_2 & 0 \\ 0 & S_1^T\widehat{K}_{21}^4 S_1 + S_2^T\widehat{K}_{21}^3 S_2 & \widehat{K}_{11}^2 + S_1^T\widehat{K}_{22}^4 S_1 + S_2^T\widehat{K}_{22}^3 S_2 & \widehat{K}_{12}^2 \\ 0 & 0 & \widehat{K}_{21}^2 & \widehat{K}_{22}^2 \end{bmatrix}$$

$$\times \begin{Bmatrix} \widehat{u}_1 \\ \widehat{u}_4 \\ \widehat{u}_7 \\ \widehat{u}_2 \end{Bmatrix} = \begin{Bmatrix} \widehat{f}_1 \\ 0 \\ 0 \\ \widehat{f}_2 \end{Bmatrix} \qquad (10.7)$$

Finally by condensing the redundant degrees of freedom associated with internal nodes 4 and 7 in Equation 10.7, the reduced equation can be obtained as

$$\widehat{\mathbf{K}}_{6\times6}^{\text{ed}} \begin{Bmatrix} \widehat{\mathbf{u}}_1 \\ \widehat{\mathbf{u}}_2 \end{Bmatrix} = \begin{Bmatrix} \widehat{\mathbf{f}}_1 \\ \widehat{\mathbf{f}}_2 \end{Bmatrix} \qquad (10.8)$$

where $\widehat{\mathbf{K}}^{\mathbf{ed}}$ is the elemental dynamic stiffness matrix with embedded de-lamination.

10.5 Damage detection and de-noising using wavelet analysis

In this chapter, a wavelet-based damage identification technique using time of arrival information is implemented for damage detection from wave propagation responses in de-laminated composite beams. Here, the responses simulated using the developed WSFE model are used to represent the experimental responses for the inverse problem of predicting damage location. The procedure involves measuring the time of arrival of waves reflected from the damage and knowing the distance traveled by the wave from the speed of the wave from the dispersion curves at the excitation frequency, the position of damage can be established. The wave speeds at required frequencies can be obtained from the dispersion relations which, as mentioned earlier, are directly available from WSFE formulation.

However, the implementation of the above technique may not be so direct in many cases. In problems, where the lengths of the waveguides are very small and if the damage is present very near to the discontinuities, there many be several reflected waves in a very short time interval. In such cases, the different reflected waves are not very differentiable and their time of arrivals cannot be measured directly from such responses. This is illustrated with several examples later in the section on numerical experiments. Wavelet analysis is used here primarily to derive, from such responses, the time of arrival of a wave generated by the defect.

Wavelet analysis or filtering gives the time-frequency map of a signal. In other words, the wavelet filtering decomposes a signal into different frequency bands or wavelet levels and within each of them, the time information is retained with a certain resolution. Here, first a fast Daubechies wavelet transform [57] of the response is performed. The wavelet coefficients of the level containing the frequencies of the loading frequencies are retained. In cases where more than one of such level exists, the lower frequency level is considered for further analysis. These retained wavelet coefficients give a very good estimation of the arrival time of different waves, as has been shown through numerical experiments in the next section. Another advantage of this method is that since the higher frequency components are filtered out and not considered in the analysis, the presence of noise, which has higher frequencies, in the response does not affect the results.

It should also be mentioned that though here wavelet analysis is done for the time domain response, for model-based damage estimation methods, the

wavelet analysis can be incorporated to the WSFE solver prior to the inverse wavelet transform and will result in much computational savings. Such
model-based damage detection techniques may be required for more precise
estimation of both extent and position.

Experimentally measured responses are always associated with noise and
thus de-noising is an important issue. Here, a wavelet-based de-noising is
done to the simulated responses with added white noise. Since the de-noising
method is based on wavelet analysis it can be done along with the damage
detection procedure. The universal threshold λ used is

$$\lambda = \sigma\sqrt{2\log(n)} \tag{10.9}$$

where σ is the standard deviation and n is the length of the signal. The soft
thresholding [23] to wavelet coefficients d_i at certain level is done as follows

$$d_i \leftarrow d_i \ \text{ for, } \ d_i < \lambda$$
$$d_i \leftarrow sign(d_i)(|d_i| - \lambda) \ \text{ for, } \ d_i \geq \lambda \tag{10.10}$$

Similar to the wavelet-based damage detection, for simulation of responses
from noisy experimentally obtained loading or excitations, de-noising can be
done after forward wavelet transform. Thus, this wavelet-based de-noising
can also be included in the WSFE solver without much increase in the computational cost.

10.6   Wave propagation in delaminated composite beam and damage detection

Here, first the formulated WSFE model of de-laminated composite beam
is used to simulate responses for different fixed-free de-laminated AS4/3501-
6 graphite-epoxy beams. The material properties of the beam are given in
Table 5.2. The beam configuration is shown in Figure 10.8, where L and
L_d are the lengths of the beam and de-lamination respectively and L_1 is the
distance of the de-lamination from the free end of the beam. The cross-
sectional dimension is 0.01×0.01 m^2 with depth $2h = 0.01$ m and width $2b = 0.01$ m. Numerical examples are presented for different values of these lengths,
ply-layup sequences, and positions of de-laminations along the thickness of
the beam. In addition, the wave propagation responses are studied for both
broad-banded impulse and narrow-banded modulated pulse loadings. The
unit impulse load used in the previous examples has a duration of 50 μs
and a frequency content of 44 kHz. Similarly, the modulated pulse loading
with central frequency of 70 kHz is shown in time and frequency domains in
Figure 10.9.

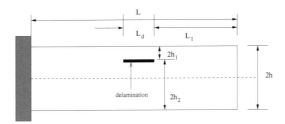

FIGURE 10.8: Fixed-free beam configuration with mid-plane delamination

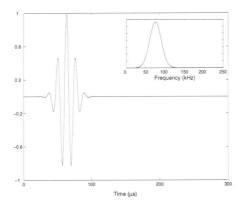

FIGURE 10.9: Narrow-banded pulse modulated at 70 kHz in time and frequency (inset) domain

Next, the responses of the de-laminated beams, simulated with two- dimensional FE, are used for the inverse problem of damage detection using wavelet analysis. As discussed in the previous section, the position of de-lamination is estimated from the arrival time of waves reflected from de-lamination. Numerical experiments are performed to predict positions of damages from simulated responses, where these responses contain artificially added noises, which are de-noised using wavelet filtering.

10.6.0.1 Responses of de-laminated beams

First, the WSFE model of de-laminated beam is validated with responses simulated using two-dimensional FE. The transverse tip velocity of beam shown in Figure 10.8 with ply-layup $[0]_8$, $L = 0.5$ m, and $L_1 = 0.25$ m is plotted for centerline de-lamination length of $L_d = 20$ mm and compared with the response obtained using two-dimensional FE in Figure 10.10(a). The FE result is obtained using 400, 4-noded quadrilateral plane stress elements and Newmark's time integration with time step 1 μs. It can be seen that results compare well. However, the small difference in the wave speeds predicted by the two methods can be further reduced by refining the FE mesh.

In Figure 10.10(b), the transverse tip velocities of same beam configuration are plotted, but for different lengths, L_d, of de-laminations and are compared with the undamaged response. The de-laminations are along the centerline of the beam and are of lengths $L_d = 10$, 20, and 30 mm. The excitation is the unit impulse load applied at tip in the transverse direction. It can be seen that in addition to the reflection from the fixed end, the damaged responses show early reflections generated from the de-laminations and amplitudes of these reflected waves increase with increase in de-lamination lengths, as expected. Similar tip transverse velocities are presented in Figure 10.10(c), except that, here, the de-lamination length is kept fixed at $L_d = 20$ mm while the positions along the thickness direction are varied from $h_1 = h$, $h/2$ and $h/4$. As in the previous plot, even here the damaged responses show reflections from the de-laminations and it can be observed that their amplitudes increase as h_1, i.e., depth of the de-lamination from the top surface of the beam, decreases. In Figure 10.10(d), the transverse velocities due to tip transverse impulse load are plotted for beams with different ply orientation sequences. In all the cases, the beam configuration is similar to Figure 10.8 with $L = 0.5$ m, $L_1 = 0.25$ m, and centerline de-lamination of $L_d = 20$ mm. The three ply-layups used are $[0]_8$, $[45]_8$, and $[60]_8$. Different ply-layups change the stiffness of the beam and hence the wave speeds also change as can be seen from Figure 10.10(d), where the responses show different amplitudes and time of arrival of reflections. The $[0]_8$ beam has the lowest amplitude and the time of arrival as it has the highest flexural stiffness and hence highest group speed.

Next, numerical experiments are performed using narrow-banded sinusoidal load (see Figure 10.9) with central frequency of 50 kHz as input excitation. For such loadings, the waves traverse non-dispersively and are widely used

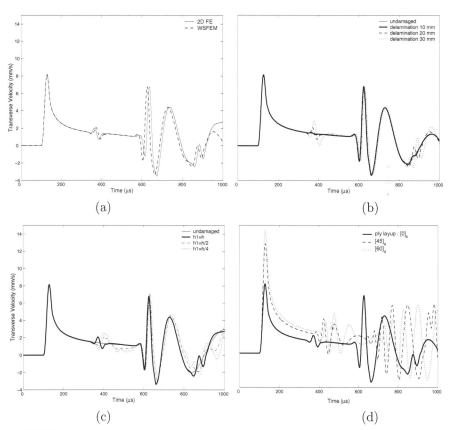

FIGURE 10.10:   Transverse tip velocity of fixed-free graphite-epoxy beam due to tip impulse load applied in transverse direction with (a) de-lamination length L_d = 20 mm (validated with 2-D FE), (b) different de-lamination lengths L_d = 10, 20, 30 mm, (c) de-laminations at different heights above centreline and (d) different ply-layup $[0]_8$, $[45]_8$, $[60]_8$

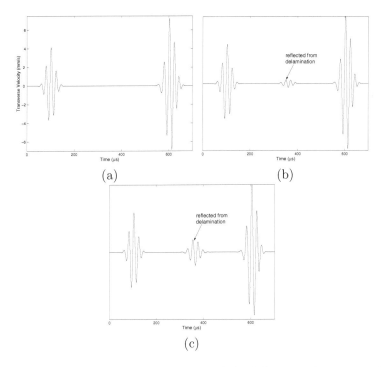

FIGURE 10.11: Transverse velocities of fixed-free graphite-epoxy beam due to narrow-banded load at 50 kHz applied in transverse direction (a) undamaged, (b) de-laminated $L_d = 10$ mm, and (c) $L_d = 20$ mm

for damage detection applications. The load is again applied in transverse direction at the tip of a $[0]_8$ beam shown in Figure 10.8 with $L = 0.5$ m and $L_1 = 0.25$ m. The responses studied are the transverse velocities measured at the tip. In Figures 10.11(a) to (c), the velocities of undamaged and de-laminated beams with $L_d = 10$ and 20 mm are plotted respectively. In either cases the de-lamination is along the centerline of the beam. Similar to the responses due to impulse loading in previous examples, the damaged responses here show an additional reflection from the de-lamination and their amplitude increases with increase in the de-lamination length. From these plots, the positions of damages can be obtained directly using the time of arrival and wave speed.

The wave speeds in terms of group speeds at a required frequency can be obtained from the dispersion plot shown in Figure 10.12 for the given $[0]_8$ graphite-epoxy beam. In this figure, the normalized group speeds C_g/C_0 are presented, where C_0 is the axial wave speed given by $C_0 = \sqrt{EA/\rho A}$, E, A, and ρ being the Young's modulus, cross-sectional area, and mass density respectively. In addition to the transverse bending mode, the dispersion plot

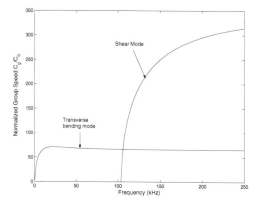

FIGURE 10.12: Dispersion relation of graphite-epoxy $[0]_8$ beam

shows the shear mode which propagates only after the cut-off frequency of approximately 105 kHz. The flexural wave speed at 50 kHz is approximately 2.08×10^3 m/s. In Figures 10.11(b) and (c), the time of arrival of wave reflected from de-lamination is nearly 0.263 ms. The position of the de-lamination predicted from using the time of arrival and wave speeds is 0.273 m from the tip, which is very near to the actual distance of 0.25 m.

In Figure 10.13(a), the response shown in Figure 10.11(b) is plotted up to 1.0 ms and in Figure 10.13(b), the corresponding noisy signal with signal to noise ratio, SNR = 30 obtained by artificially adding white noise, is presented. It can be seen that, in the noisy signal, the waves reflected from de-lamination are not distinguishable visually. The de-noised signal is shown in Figure 10.13(c) and is obtained through universal soft thresholding as described earlier. The thresholding is done only in the first three wavelet levels corresponding to higher frequencies and order of Daubechies basis used for wavelet decomposition is $N = 22$. Though the first reflected wave from the de-lamination is slightly distorted in the de-noised response, the second reflection from de-lamination can be clearly identified.

Figures 10.14(a) to (c) show the tip transverse velocities due to different positions of de-lamination with $L_d = 20$ mm along the centerline, i.e., $L_1 = 0.2, 0.3,$ and 0.4 m, respectively. The beam configuration is otherwise similar to that used in the previous example. The loading is a narrow-banded pulse (see Figure 10.9) modulated at 50 kHz and is applied at the tip in transverse direction. Here, the distances of damages from the tip predicted using time of arrival and wave speed are 0.23, 0.33, and 0.44 m, which match well with the actual positions $(L_1 + L_d/2)$ 0.21, 0.31, and 0.41 m, respectively.

It should be mentioned here, that though in the above examples with narrow-banded modulated excitation, the position of the de-lamination can be obtained directly by measuring the time of arrivals from the plots, it may not be so straightforward in many other cases. Particularly, in problems dealing with short length structures or where the damages are very near to bound-

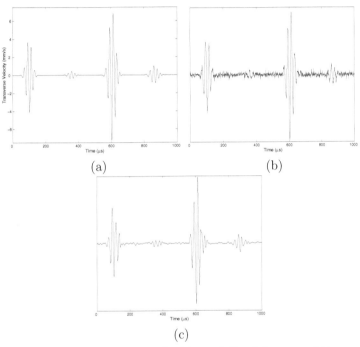

FIGURE 10.13: Transverse velocities of fixed-free graphite-epoxy delaminated ($L_d = 10$ mm) beam due to narrow-banded load at 50 kHz applied in transverse direction (a) simulated response, (b) noisy response, and (c) de-noised response

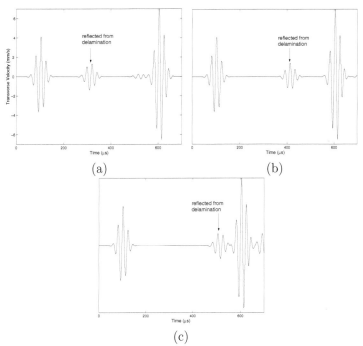

FIGURE 10.14: Transverse velocities of fixed-free graphite-epoxy delaminated ($L_d = 20$ mm) beam due to narrow-banded load at 50 kHz applied in transverse direction (a) $L_1 = 20$mm, (b) $L_1 = 30$mm, and (c) $L_1 = 40$mm (see Figure 10.8)

aries/discontinuities, the incident waves, waves reflected from damages, and boundaries/discontinuities are cluttered together. In such situations, it is very difficult to extract the time of arrivals directly from the plots. Wavelet analysis of measured response helps to alleviate this problem to a great extent. This is explained with numerical examples in the next section.

10.6.0.2   Damage detection using wavelet analysis

A wavelet-based damage detection technique as discussed earlier is adopted here for predicting the positions of de-laminations. First, responses are simulated for a $[0]_8$ graphite-epoxy beam with embedded de-lamination of $L_d = 20$ mm along the centerline. Simulations are done using two-dimensional FE, with 400, 4-noded quadrilateral plane stress elements and Newmark's time integration with time step 1 μs. The beam is similar to that shown in Figure 10.8 and used in previous experiments except that the length L is shorter and is equal to 0.15 m. The excitation is the narrow-banded pulse with central frequency 70 kHz shown in Figure 10.9 applied at the tip in transverse direction. In Figures 10.15(a) and (b), the transverse tip velocities of the undamaged and de-laminated beams with de-lamination at $L_1 = 0.04$ m from the tip are presented respectively. It can be seen that in the damaged response shown in Figure 10.15(b), the incident wave and wave reflected from de-lamination are not distinguishable. Similarly, in Figure 10.15(c), the response of de-laminated beam with de-lamination at $L_1 = 0.09$ m from tip, i.e., near to the fixed end of the beam, is presented. Again in this case, the waves reflected from the de-lamination and the fixed boundary are cluttered together.

To determine the positions of damages from these simulated responses, first wavelet analysis is done using Daubechies basis of order $N = 22$. This generates wavelet coefficients of the responses at different frequency bands or wavelet levels. Here, the wavelet level corresponding to the highest frequency band is referred to as level one. Since the frequency content of the applied load with central frequency of 70 kHz is 20 to 120 kHz, the responses will also have the same frequency band. Thus, wavelet decomposition is done only up to first three levels corresponding to frequencies 125 to 250 kHz, 62.5 to 125 kHz, and 31.25 to 62.5 kHz respectively. In Figure 10.16(a), the normalized absolute values of the wavelet coefficients at level three are plotted for the undamaged and damaged responses shown in Figures 10.15(a) and (b). The plots show few peaks which represent different waves. The wavelet coefficients of the damaged response show two additional peaks due to the waves reflected from de-lamination. The time of arrivals of the incident wave and the wave reflected from the fixed end are $t_i = 48$ μs, $t_r = 192$ μs respectively for both undamaged and damaged responses and they match exactly. The two additional peaks in the damaged responses arrive at $t_{d1} = 96$ μs and $t_{d2} = 240$ μs. The speed of the flexural mode at 50 kHz derived from the dispersion relation presented in Figure 10.12 as mentioned before is 2.08×10^3 m/s. This speed is used to

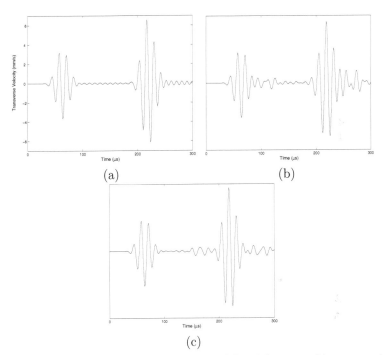

FIGURE 10.15: Transverse velocities of fixed-free graphite-epoxy beam due to narrow-banded load at 70 kHz applied in transverse direction (a) undamaged, (b) de-laminated $L_d = 20$ mm and $L_1 = 40$, and (c) de-laminated $L_d = 20$ mm and $L_1 = 90$

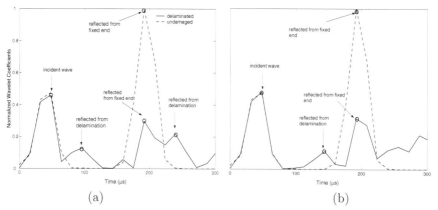

(a)   (b)

FIGURE 10.16:   Normalized wavelet coefficients of level corresponding to frequency content 31.25 – 62.5 kHz of undamaged and damaged responses (a) Figures 10.15(a) and (b), respectively and (b) Figures 10.15(a) and (c), respectively

calculate the position of de-lamination from the arrival time since 50 kHz is approximately the average of the frequency band in wavelet level three. Using this speed, the distance of the fixed end predicted from t_i and t_r is 0.15 m which is exactly equal to that used in simulations of the responses. Similarly, the position of the damage obtained using both t_i, t_{d1} and t_i, t_{d2} is 0.05 m from the tip which matches exactly with actual distance of 0.05 $(= L_1 + L_d/2)$ m.

Similar wavelet coefficients for undamaged and damaged responses shown in Figures 10.15(a) and (c) respectively are presented in Figure 10.16(b). Even in this case, the arrival times of incident wave and wave reflected from the fixed end are $t_i = 48$ μs and $t_r = 192$ μs respectively, which are exactly similar to that obtained in the previous example. The time of arrival of the wave resulted from de-lamination is $t_{d1} = 144$ μs and thus, using the wave speed at 50 kHz, the predicted position of the damage from the tip is 0.1 m, which is the actual distance used in simulation.

It should be mentioned here that similar analysis could have been done using the coefficients corresponding to the wavelet level two with frequency content 62.5 to 125 kHz. However, the level with lower frequencies is used for analysis with the motive to eliminate effects of noises which are generally of higher frequency content and are very often present in experimentally measured responses.

Next, similar wavelet analysis is done for prediction of damage location from noisy response. The noisy response presented in Figure 10.17(a) is obtained by adding white noise to the response shown in Figure 10.15(b) to produce a SNR = 15. In Figure 10.17(b), the corresponding de-noised response is plotted. The wavelet-based de-noising, described earlier, is done at the first

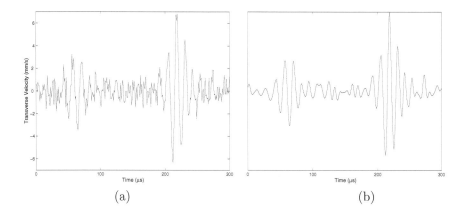

FIGURE 10.17: Transverse velocities of fixed-free graphite-epoxy de-laminated beam due to narrow-banded load at 70 kHz applied in transverse direction, shown in Figure 10.15(b), (a) noisy response and (b) de-noised response

two wavelet levels and the order of Daubechies basis used is $N = 22$. In Figure 10.18, the normalized absolute wavelet coefficients at level three are plotted for the undamaged response, noisy damaged response and the corresponding de-noised responses. The arrival times t_i, t_r, t_{d1}, and t_{d2} in this case are exactly similar to those in the previous example shown in Figure 10.16(a) for the response without any added noise. In addition, it can be seen that the coefficients for the noisy and de-noised responses overlap. These two observations indicate that in the wavelet-based damage detection technique, the presence of noise in the response does not affect the prediction of damage location. This can be explained as the wavelet coefficients used for analysis have low frequency content and the higher frequency contents including noises are filtered out.

This final chapter of the book shows the implementation of the WSFE method to solve inverse problems of force reconstruction and damage detections which are important applications of wave propagation as mentioned earlier. Apart from the use of WSFE for simulations, wavelet is also used here for the purpose of signal processing to extract damage features from the wave response of the damaged structures.

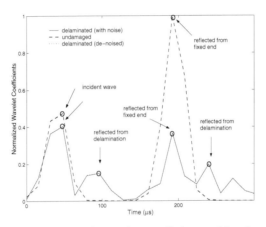

FIGURE 10.18:   Normalized wavelet coefficients of level corresponding to frequency content 31.25 – 62.5 kHz of undamaged, noisy, and de-noised damaged responses (Figures 10.15(a), 10.17(a) and (b), respectively)

References

[1] K. Amaratunga and J. R. Williams. Time integration using wavelets. pages 894–902. Proceedings of SPIE, Wavelet Application for Dual Use. 2491, Orlando, FL, 1995.

[2] K. Amaratunga and J. R. Williams. Wavelet-Galerkin solution of boundary value problems. *Archives of Computational Methods in Engineering*, 4(3):243–285, 1997.

[3] K. J. Bathe. Finite element procedures. Prentice Hall, NJ, 1996.

[4] R. C. Batra and H. K. Ching. Analysis of elasto dynamic deformations near a crack/notch tip by the meshless local Petrov-Galerkin (MLPG) method. *CMES: Computer Modeling in Engineering and Sciences*, 3(6):717–730, 2002.

[5] R. H. Baughman, C. X. Cui, A. A. Zakhidov, Z. Iqbal, J. N. Barisci, and G. M Spinks. Carbon nanotube actuators. *Science*, 284:1340–1344, 1999.

[6] S. Bertoluzza and G. Naldi. Some remarks on wavelet interpolation. *Computational and Applied Mathematics*, 13(1):13–32, 1994.

[7] D. E. Beskos. Boundary element method in dynamic analysis Part II (1986–1996). *Applied Mechanics Review*, 50:149–197, 1997.

[8] G. Beylkin. On the representation of operators in bases of compactly supported wavelets. *SIAM Journal of Numerical Analysis*, 6(6):1716–1740, 1992.

[9] A. Chakraborty and S. Gopalakrishnan. Force identification in a fgm plate using truncated response. 44th AIAA/ASME/ASCE/AHS-SDM Conference, Paper No. AIAA 2003-1590, Norfolk, Virginia, 2003.

[10] A. Chakraborty and S. Gopalakrishnan. Wave propagation in inhomogeneous layered media: Solution of forward and inverse problems. *Acta Mechanica*, 169:153–185, 2004.

[11] A. Chakraborty, M. S. Sivakumar, and S. Gopalakrishnan. Spectral element based model for wave propagation analysis in multi-wall carbon nanotubes. *International Journal of Solids and Structures*, 43(2):279–294, 2006.

[12] T. Chang and H. Gao. Size dependence elastic properties of a single-walled carbon nanotube via a molecular mechanics model. *Journal of Mechanics and Physics of Solids*, 51:1059, 2003.

[13] M. Q. Chen, C. Hwang, and Y. P. Shih. The computation of wavelet-Galerkin approximation on a bounded interval. *International Journal for Numerical Methods in Engineering*, 39:2921–2944, 1996.

[14] L. Chico, R. Pérez-Álvarez, and C. Cabrillo. Continuum model for long-wavelength optical phonons in cylinders: Application to carbon nanotubes. *Physical Review B*, 69:035419, 1–6, 2004.

[15] L. Chico, R. Pérez-Álvarez, and C. Cabrillo. Low frequency phonons in carbon nanotubes: A continuum approach. *Physical Review B*, 73:075425, 1–7, 2006.

[16] A. Cohen. Biorthoganal wavelets. pages 123–152. Wavelets: A tutorial in theory and applications. C. K. Chui, ed., Academic, San Diego, 1992.

[17] A. Cohen, I. Daubechies, and J. C. Feauvean. Biorthogonal bases of compactly supported wavelets. *Communications in Pure and Applied Mechanics*, 45:485–560, 1992.

[18] R. D. Cook, D. S. Malkus, M. E. Plesha, and R. J. Witt. Concepts and applications of finite element analysis. Wiley, NY, 2003.

[19] I. Daubechies. Orthonormal bases of compactly supported wavelets. *Comm. in Pure and Applied Mathematics*, 41:906–966, 1988.

[20] I. Daubechies. Ten lectures on wavelets. CBMS-NSF Series in Applied Mathematics (SIAM, Philadelphia), 1992.

[21] P. J. Davis. Interpolation and approximation. Blaisdell, New York, 1963.

[22] D. L. Donoho. Interpolating wavelet transforms. Technical Report, Department of Statistics, Stanford University, 1992.

[23] D. L. Donoho. De-noising by soft thresholding. *IEEE Transactions on Information Theory*, 41:613–627, 1995.

[24] J. F. Doyle. Further developments in determining the dynamic contact law. *Experimental Mechanics*, 24:265–270, 1984.

[25] J. F. Doyle. Determining the contact force during transverse impact of plates. *Experimental Mechanics*, 27:68–72, 1987.

[26] J. F. Doyle. Experimentally determining the contact force during the transverse impact of ortho-tropic plates. *Journal of Sound and Vibration*, 118(3):441–448, 1987.

[27] J. F. Doyle. Force identification from dynamic response of a bi-material beam. *Experimental Mechanics*, 33:64–69, 1993.

[28] J. F. Doyle. Wave propagation in structures. Springer, New York, 1999.

[29] J. F. Doyle. Reconstructing dynamic events from time limited spatially distributed data. *International Journal for Numerical Methods in Engineering*, 53:2721–2734, 2002.

[30] M. S. Dresselhaus and P. C. Eklund. Phonons in carbon nanotubes. *Advances in Physics*, 49:705–814, 2000.

[31] L. Elden, F. Berntsson, and T. Reginska. Wavelet and fourier methods for solving the sideways heat equation. *SIAM Journal of Scientific Computing*, 21(6):2187–2205, 2000.

[32] A. M. Fennimore, T. D. Yuzvinsky, W. Q. Han, M. S. Fuhrer, J. Cumings, and A. Zetti. Rotational actuators based on carbon nanotubes. *Nature*, 424:408–410, 2003.

[33] S. J. V. Frankland and V. M. Harik. Analysis of carbon nanotube pull-out from a polymer matrix. NASA/CR-2002211743, ICASE Report No. 2002-23, 2002.

[34] X.-L. Gao and K. Li. A shear-lag model for carbon nanotube-reinforced polymer composites. *International Journal of Solids and Structures*, 42:1777–1795, 2005.

[35] R. F. Gibson, E. O. Ayorinde, and Y. F. Wen. Vibrations of carbon nanotubes and their composites: A review. *Composite Science and Technology*, 67:1–28, 2007.

[36] S. Gopalakrishnan, A. Chakraborty, and D. Roy Mahapatra. Spectral finite element method: Wave propagation, diagnostics and control in anisotropic and inhomogeneous structures. Springer, 2007.

[37] K. F. Graff. Wave motions in elastic solids. Ohio State University Press, 1975.

[38] A. Grossmann and J. Morlet. Decomposition of Hardy functions into square integrable wavelets of constant shape. *SIAM Journal of Mathematical Analysis*, 15:723–736, 1984.

[39] J. H. Heimann and K. J. Kolsky. The propagation of elastic waves in thin cylindrical shells. *Journal of Mechanics and Physics of Solids*, 14:121–130, 1966.

[40] F. Q. Hu, M. Y. Hussaini, and P. Rasetarinera. An analysis of the discontinuous Galerkin method for wave propagation methods. *Journal of Computational Physics*, 151(2):921–946, 1999.

[41] S. Ijima. Helical microtubules of graphitic carbon. *Nature*, 354:56–58, 1991.

[42] R. A. Jishi, L. Venkataraman, M. S. Dresselhaus, and G. Dresselhaus. Phonon modes in carbon nanotubules. *Chemical Physics Letters*, 209:77–82, 1993.

[43] A. Jorio, M. S. Dresselhaus, and G Dresselhaus. Carbon nanotubes: Advanced topics in the synthesis, structure, properties and applications. Springer, 2008.

[44] P. Lancaster. Lambda matrices and vibrating systems. Pergamon Press, 1966.

[45] P. Lancaster. Theory of matrices. Academic Press, 1969.

[46] A. Latto, H. Resnikoff, and E. Tanenbaum. The evaluation of connection coefficients of compactly supported wavelets. Proceedings of French-USA Workshop on Wavelets and Turbulence, Princeton University, June 1991, Springer, New York, 1992.

[47] C. Li and T. W. Chou. Multiscale modeling of carbon nanotube reinforced polymer composites. *Journal of Nanoscience and Nanotechnology*, 3:423, 2003.

[48] C. Li and T. W. Chou. Elastic wave velocities in single-walled carbon nanotubes. *Physical Review B*, 73:245407, 1–5, 2006.

[49] C. Li and T. W. Chow. Vibrational behaviors of multiwalled-carbon-nanotube-based nanomechanical resonators. *Applied Physics Letters*, 84:121–123, 2004.

[50] C. Li, E. T. Thostenson, and T. W. Chou. Sensors and actuators based on carbon nanotubes and their composites: A review. *Composites Science and Technology*, 68(6):1227–1249, 2008.

[51] C. Y. Li and T. W. Chou. Strain and pressure sensing using single-walled carbon nanotubes. *Nanotechnology*, 15:1493–1496, 2004.

[52] Z. Li, S. Xia, J. Fau, and X. Su. Damage detection of beams based on experimental wavelet analysis of flexible waves. *Key Engineering Materials*, 261:1373–1378, 2004.

[53] A. E. H. Love. A treatise on the mathematical theory of elasticity. Dover, New York., 1944.

[54] J. P. Lu. Elastic properties of carbon nanotubes and nanoropes. *Physical Review Letters*, 79:1297, 1997.

[55] G. D. Mahan. Oscillations of thin hollow cylinders: Carbon nanotubes. *Physical Review B.*, 65:235404 1–7, 2002.

[56] D. Roy Mahapatra and S. Gopalakrishnan. A spectral finite element model for analysis of axial flexural shear coupled wave propagation in laminated composite beams. *Composite Structures*, 59(1):67–88, 2003.

[57] M. Mallat. A theory for multiresolution signal decomposition; the wavelet representation. *IEEE Trans. on Pattern Analysis and Machine Intelligence*, 11(7):659–674, 1988.

[58] S. G. Mallat. Multiresolution approximation and wavelets. Preprint GRASP Lab., Department of Computer and Information Science. University of Pennsylvania, 1986.

[59] W. J. Mansur. A time stepping technique to solve wave propagation problems using the boundary element method. PhD Thesis, Southampton University, 1983.

[60] L. Meirovitch. Fundamentals of vibrations. McGraw-Hill, New York, 2002.

[61] Y. Meyer. Principe d'incertitude, bases hilbertiennes et algébres d'operateurs'. volume 662. Seminaire Bourbaki, 1987.

[62] M. Mitra and S. Gopalakrishnan. Spectrally formulated wavelet finite element for wave propogation and impact force identification in connected 1-D waveguides. *International Journal of Solids and Structures*, 42:4695–4721, 2005.

[63] M. Mitra and S. Gopalakrishnan. Extraction of wave charateristics from wavelet based spectral finite element formulation. *Mechanical Systems and Signal Processing*, 20:2046–2079, 2006.

[64] M. Mitra and S. Gopalakrishnan. Wave propagation analysis in carbon nanotube embedded composite using wavelet based spectral finite elements. *Smart Materials and Structures*, 15:104–122, 2006.

[65] M. Mitra and S. Gopalakrishnan. Wavelet based spectral finite element for analysis of coupled wave propagation in higher order composite beams. *Composite Structures*, 73:263–277, 2006.

[66] M. Mitra and S. Gopalakrishnan. Vibrational characteristics of single-walled carbon-nanotube: Time and frequency domain analysis. *Journal of Applied Physics*, 101:114320, 2007.

[67] M. Mitra and S. Gopalakrishnan. Wave propagation in imperfectly bonded single walled carbon nanotube-polymer composites. *Journal of Applied Physics*, 102:084301, 2007.

[68] Mira Mitra and S. Gopalakrishnan. A wavelet-based spectral finite element for analysis of coupled wave propagation in composite beam. pages 370–376. In Proceedings of the Third MIT Conference on Computational Fluid and Solid Mechanics, MIT, Cambridge, 2005.

[69] J. Morlet. Sampling theory and wave propagation. pages 233–261. in C. H. Chen (ed.), NATO ASI Series, Issues in Acoustic Signal/Image Processing and Recognition, Vol 1, Springer, Berlin, 1983.

[70] J. Morlet, G. Arens, I. Fourgeau, and D. Giard. Wave propagation and sampling theory. *Geophysics*, 47:203–236, 1982.

[71] A. Nag, D. R. Mahapatra, S. Gopalakrishnan, and T. S. Sankar. A spectral finite element with embedded de-lamination for modeling of wave scattering in composite beams. *Composites Science and Technology*, 63:2187–2200, 2003.

[72] G. V. Narayanan and D. E. Beskos. Use of dynamic influence coefficients in forced vibration problems with the aid of fast Fourier transform. *Computers and Structures*, 9:445–450, 1978.

[73] A. H. Nayfeh and P. F. Pai. Linear and non-linear structural analysis. Wiley, New Jersey., 2004.

[74] D. E. Newland. Wavelet analysis of vibration. Part I and Part II. *Transactions of ASME: Journal of Sound and Vibration*, 116:409–425, 1994.

[75] H. S. Park and W. K. Liu. An introduction and tutorial on multiple scale analysis in solids. *Computer Methods in Applied Mechanics and Engineering*, 193:1733–1772, 2004.

[76] S. H. Park and J. L. Tassoulas. A discontinuous Galerkin method for transient analysis of wave propagation in unbounded domains. *Computer Methods in Applied Mechanics and Engineering*, 191(36):3983–4011, 2002.

[77] R. D. Patton and P. C. Marks. One dimensional finite elements based on the Daubechies family of wavelets. *AIAA Journal*, 34(8):1696–1698, 1996.

[78] V. N. Popov, V. E. Van Doren, and M. Balkanski. Elastic properties of single-walled carbon nanotubes. *Physical Review B*, 61:003078, 2000.

[79] S. Qian and J. Weiss. Wavelets and the numerical solution of boundary value problems. *Applied Mathematics Letter*, 6(1):47–52, 1993.

[80] S. Qian and J. Weiss. Wavelets and the numerical solution of partial differential equations. *Journal of Computational Physics*, 106(1):155–175, 1993.

[81] A. M. Rao, E. Richter, S. Bandow, B. Chase, P. C. Eklund, K. A. Williams, S. Fang, K. R. Subbaswamy, M. Menon, A. Thess, R. E. Smalley, G. Dresselhaus, and M. S. Dresselhaus. Diameter-selective Raman scattering from vibrational modes in carbon nanotubes. *Science*, 275:187, 1997.

[82] T. Reginska. Sideways heat equation and wavelets. *Journal of Computational and Applied Mathematics*, 63:209–214, 1995.

[83] T. Reginska and L. Elden. Solving the sideways heat equation by a wavelet-Galerkin method. *Inverse Problems*, 13:1093–1106, 1997.

[84] S. A. Rizzi and J. F. Doyle. Force identification for impact problem on a half plane. *Computational Techniques for Contact, Impact, Penetration and Perforation of Solids*, 103:163–182, 1989.

[85] S. A. Rizzi and J. F. Doyle. Force identification for impact of a layered system. Computational Aspects of Contact, Impact and Penetration. *Elmepress International*, pages 222–241, 1991.

[86] C. Q. Ru. Column buckling of multiwalled carbon naotubes with interlayer radial displacements. *Physical Review B*, 62:16962–16967, 2000.

[87] V. Sazonova, Y. Yaish, H. Ustenel, D. Roundy, T. A. Arias, and P. McEuen. A tunable carbon nanotube elctromechanical oscillator. *Nature*, 431:284–287, 2004.

[88] B. F. Shorr. The wave finite element method. Springer, Berlin, 2004.

[89] I. Sneddon. Fourier transforms. McGraw Hill, New York., 1951.

[90] K. Sohlberg, B. G. Sumpter, R. E. Tuzun, and D. W. Noid. Continuum methods of mechanics as a simplified approach to structural engineering of nanostructures. *Nanotechnology*, 9:30–36, 1998.

[91] H. Sohn, G. Park, J. R. Wait, N. P. Limback, and C. R. Farrar. Waveletbased active sensing for delamination detection in composite structures. *Smart Materials and Structures*, 13:153–160, 2004.

[92] D. Srivastava, C. Wei, and K. Cho. Nanomechanics of carbon nanotubes and composites. *Applied Mechanics Review*, 56(2):215–230, 2003.

[93] J. O. Stromberg. A modified Franklin system and higher order spline systems on r^n as unconditional bases for hardy spaces. pages 475–493. Conf. in Honour of A. Zygmund, Vol II, Wadsworth Mathematics Series, 1982.

[94] E. T. Thostenson, C. Li, and T. W. Chou. Nano-composites in context. *Composites Science and Technology*, 65:491–516, 2005.

[95] E. T. Thostenson, Z. Ren, and T. W. Chou. Advances in science and technology of carbon nanotubes and their composites: a review. *Composites Science and Technology*, 61:1899–1912, 2001.

[96] P. Vandergheynst, J.-P. Antoine, E. Van Vyve, A. Goldberg, and I. Doghri. Modeling and simulation of an impact test using wavelets, analytical solutions and finite elements. *International Journal of Solids and Structures*, 38:5481–5508, 2001.

[97] H. D. Wagner. Nanotubepolymer adhesion: a mechanics approach. *Chemical Physics Letters*, 361:57–61, 2002.

[98] C. Y. Wang, C. Q. Ru, and A. Mioduchowski. Applicability and limitations of simplified elastic shell equations for carbon nanotubes. *Journal of Applied Mechanics*, 71:622, 2004.

[99] C. Y. Wang, C. Q. Ru, and A. Mioduchowski. Axisymmetric and beam-like vibrations of multi-wall carbon nanotubes. *Physical Review B*, 72:075414 1–10, 2005.

[100] J. R. Williams and K. Amaratunga. Introduction to wavelets in engineering. *International Journal for Numerical Methods in Engineering*, 37:2365–2388, 1994.

[101] J. R. Williams and K. Amaratunga. A discrete wavelet transform without edge effects using wavelet extrapolation. *Journal of Fourier Analysis and Applications*, 3(4):435–449, 1997.

[102] B. I. Yakobsen, C. J. Brabec, and J. Bernholc. Nanomechanics of carbon tubes: instabilities beyond linear response. *Physical Review Letters*, 76:2511, 1996.

[103] L. H. Ye, B. G. Liu, D. S. Wang, and R. Han. Ab initio phonon dispersions of single-wall carbon nanotubes. *Physical Review B*, 69:235409, 2004.

[104] J. Yoon, C. Q. Ru, and A. Mioduchowski. Sound wave propagation in multiwall carbon nanotubes. *Journal of Applied Physics*, 93:4801–4806, 2003.

[105] J. Yu, R. K. Kalia, and P. Vashishta. Phonons in graphitic tubules: A tight-binding molecular dynamics study. *Journal of Chemical Physics*, 103:6697, 1995.

[106] S. Yuan, W. Lei, and L. Shi. Active monitoring for on-line damage detection in composite structures. *Journal of Vibration and Acoustics*, 125:178–186, 2003.

[107] Q. Zhang and Q. Jiang. Multiwalled carbon nanotubes as gigahertz oscillators. *Physical Review Letters*, 88:045503–1–3–3, 2002.

Index

Approximation
    bases, 1
    coefficients, 19, 26, 50, 82
    function, 49
    spatial, 141, 146
    temporal, 141, 144
Atomistic simulation, 184, 208
Axisymmetric cylinder, 165, 185

Basis function, 5, 19
Bessel function, 164
Bi-material cylinder, 141
Boundary
    coefficient, 83
    element, 4

Carbon nanotube pull-out, 208
Central difference, 3
CFT, 14
Characteristic equation, 64, 66
Circulant matrix, 118, 119, 148
Circular convolution method, 51
Classical plate theory, 141
Composite
    anisotropic, 141
    beam, 101, 127
    graphite-epoxy, 108, 160
Connection coefficients, 51, 55, 83
Continuous system, 38, 62
Convolution, 12, 119
    integral, 46, 48, 60
Coupled, 4
    axial-radial, 235
    differential equations, 3
    radial-longitudinal mode, 189
    vibration, 189
Cut-off frequency, 66, 112, 137, 218

Damage detection, 37, 247, 251
Damping mode, 63, 67
Damping ratio, 40
De-lamination, 247, 248
De-noising, 1, 37, 242, 251
DFT, 14, 51, 63, 64, 66
Diagnostic wave, 247
Dilation, 19
Dispersion relation, 63, 120, 185, 218
Dispersive, 63, 129, 241
Dyadic, 23
Dynamic stiffness matrix, 76, 85, 95

Eigenvalue, 61, 70
    analysis, 2, 3, 83, 147
    problem, 23
Eigenvector, 70, 118
    matrix, 3, 61, 84
EOM, 47, 60, 62
Equation of motion, 38, *see* EOM
Euler-Bernoulli beam, 66, 91, 123
Evanescent mode, 67, 126

Fast Fourier transform, 5, *see* FFT
FE
    method, 4, 71, 242
    modeling, 4
FFT, 72, 120
Filter
    coefficient, 19, 21, 23
    high pass, 28
    low pass, 28
    quadrature mirror, 20
Finite
    domain, 5
Finite element, *see* FE

solution, 87
technique, 2
Flügge's shell model, 192, 193
Folded plate, 141, 161
Force reconstruction, 241, 243
Fourier
    series, 13, 45
    transform, 10, 50, 63, 68, 84, 89
        continuous, 5, 11
        discrete, 13
        short term, 16
        windowed, 14
Fourier transform-based SFE, 73, *see* SFE, 241
Frame structure, 98, 245
Free vibration, 38, 39, 57
Frequency
    band, 251
    bandwidth, 16
    component, 23
    content, 23
    domain, 5
        analysis, 5, 117, 155
        resolution, 5
    Nyquist, 12, 117, 120, 129
Frequency response function, 72
FSFE, 73, 76, 112, 138, 156

Graphene, 183
Group speed, 65, 67, 196

Hamilton's principle, 38, 143
Harmonic loading, 43

Impulse
    force, 243
    load, 87, 95, 109, 156, 200, 225
    response, 47
    unit, 47
Infinite beam, 129
Inverse
    transform, 13, 62

Kirchoff shear, 144

Laplace transform, 1, 9, 50, 55
Layer-wise shear deformation, 208, 209
Linear
    operation, 10
    transformation, 3, 11

Mallat's
    algorithm, 26
    decomposition, 26
    reconstruction, 28
    transform, 26
MDOF, 2, 3, 38, 50, 55, 58–62
Modal
    analysis, 2
Mode
    shape, 2, 57
    superposition, 3, 4
Multi-degree of freedom, 2, *see* MDOF, 38, 57
Multi-scale modeling, 208
Multi-walled carbon nanotube, 192, *see* MWNTs
MWNTs, 200, 209

Natural frequencies, 2
Newmark
    time integration, 3, 49
Noise, 4, 243
Non-dispersive, 63, 108, 112, 129
Non-periodic
    boundary condition, 53
    WSFE formulation, 133
Normal mode, 57

Operator, 17
Oscillators, 6

Partial stress transfer, 208, 231, 235
Penalty matrix, 208, 232
PEP, 70
Periodic
    boundary condition, 119
    excitation, 45
    extension, 147

loading, 51
solution, 119
WSFE formulation, 120
Phase speed, 65, 67
Ply-dropped plate, 161
Polynomial eigenvalue problem, 68,
  see PEP
Propagating mode, 63, 67
Pulse
  broad-banded, 108
  modulated sinusoidal, 129
  narrow-banded, 108, 229
  rectangular, 10, 11

Radial breathing mode, 185
Rectangular
  window, 16
Representative volume element, 208
Resolution, 5
  level, 26
  multi-, 17, 18
Restraint boundary condition, 148
Riesz bases, 18, 19

Scaling, 11, 19
  function, 18, 22, 51, 82
    Daubechies, 22, 50, 61
    Haar, 23
    moment, 53
SDOF, 38, 39, 41, 47, 50, 55, 61
Sensors, 6, 207
SFE, 5, 63, 71, 184
Signal processing, 1
Single-degree of freedom, 38, *see*
  SDOF
Single-walled carbon nanotube, 183,
  see SWNTs, 185
Singular value decomposition, 68,
  see SVD
Spectral
  analysis, 1, 63, 68, 120
  density, 14
  finite element, 5, *see* SFE, 89
Spectrum relation, 63, 120, 218
Steady-state response, 44

STFT, 17
Structural health monitoring, 2, 4
Subspace, 18, 19
SVD, 69
SWNTs, 198, 208
  pull-out, 229

Time
  -frequency map, 251
  approximation, 53
  integration, 4, 49, 50, 62
  resolution, 23
Timoshenko beam, 73, 123, 218,
  248
Tone-burst signal, 108
Transient
  dynamic, 4, 71
  loadings, 4
  response, 44

Un-restraint boundary condition, 147

van der Waals force, 183, 208

Wave
  amplitude, 68, 77
  mode, 63, 66
  speed, 63, 65, 117, 120
Waveguide, 6, 20, 55, 63, 66, 122
  one-dimensional, 183
  two-dimensional, 141, 183
Wavelet
  basis function, 18, 68
  bi-orthogonal, 5
  coefficients, 260
  compactly supported, 20, 51
  Daubechies, 20, 51, 53, 54
    order, 22
    orthogonal, 5
  extrapolation technique, 53, 117
  Haar, 17, 23
  interpolation, 5
  level, 251
  orthogonal, 20
  transform, 17

continuous, 17
discrete, 23
Wavenumber, 63, 68, 117, 120, 196
Windowing function, 16, 18
Wrap around, 72